# Shifting Sands

The Rise and Fall of the Glaven Ports

Godfrey Sayers

## GRATITUDE

There are two people without whom this book would never have seen the light of day. John Bond of whitefox, whose support and generosity has ensured that everything I hoped for with this book has been realised. And John Wright, whose historical knowledge and wise head have been an invaluable guide.

First published in 2021
Reprinted in 2021, 2023

Copyright © Godfrey Sayers, 2021

The moral right of Godfrey Sayers to be identified as the author of this work has been asserted in accordance with the Copyright, Designs and Patents Act 1988. All rights reserved. No part of this publication may be reproduced, stored in a retrieval system or transmitted in any form or by any means, electronic, mechanical, photocopying, recording or otherwise, without prior written permission of Godfrey Sayers.

ISBN 978-1-912892-99-0

Cover design by Tom Cabot/ketchup
Page design and layout by Tom Cabot/ketchup

Project management by whitefox
Printed and bound in Great Britain

# Contents

Introduction   5

1   1 CE   Coming in From Sea   10
2   1250   Delivering Around the Haven   24
3   Concluding the Deliveries   41
4   1586   Flemish Visitors   58
5   1586   Hansa Meeting   76
6   1845   Coastal Harbours Commissioner's Visit   90
7   1945   A Blakeney Child's Memories   108
8   The Present Day   123
9   What Might Have Been?   141

# Introduction

As an intro to a recent lecture to UEA students I said this: 'I don't have a prefix; I'm not a prof or a doc, and the letters I have after my name have absolutely nothing to do with geology, hydrology, or geomorphology; should I need a prefix then a suitable one in this context might be 'old'. I have lived here a long time, I have observed and recorded the changes; the storms, the broken banks, the retreating shoreline, growing marsh, and silting creeks, and remain insatiably curious as to how this fluid and often tempestuous world I live in functions'.

The events that drive change are random and unpredictable, but over time their consequences are quantitative, making it possible to extrapolate back in time and, although I won't be doing that here, forward into the future. Here I have tried to illustrate 2,000 years of geomorphic change and how that change directly and indirectly wrote the histories of Salthouse and the villages of the Glaven Valley. Histories that are rich and vibrant, albeit not as rich and vibrant as they might have been if man had been able to resist the urge to interfere. This book attempts to show how inextricably these histories have been tied to a dramatically changing environment. Shifting sands transported by wind, tide and fluctuating sea levels have shape-shifted the north coast of Norfolk as the wind shapes clouds. When Norwich Cathedral was being built Blakeney Point was an embryonic bump, their originations coincident. However, because geophysical change has been so rapid the rear-view mirror can get a little fogged when one tries to envision how it might have looked 2,000 years ago. We know that 9,000 years ago today's shoreline was many miles from the sea. To the north, where Cley Beach is today would have been flat grasslands with occasional low hills of glacial deposits (Eyes) similar to the Great Eye and Gramborough Hill, and the Glaven would have meandered wide and shallow across these open plains of grass and woodland to join the Stiffkey and eventually the Ouse, which finally emptied into a wide inlet behind the Dogger Hills, where vast offshore wind farms are being built today.

So where are the windows that will allow us to look back? Well, most of them can be found all around us, there are good clues in the landscape, and anecdotal evidence (normally a complete non sequitur for scientists) can, in the memories of local people, be very useful; pointing out how channels have moved, how much saltmarsh accretion has occurred during a lifetime, or how far the shingle ridge has rolled back and how often. Supported by many clues in the existing marshes and beachscapes, this can paint a detailed and reliable picture.

Along the north edge of the marsh at Blakeney, for example, is a raised ridge running from just west of Blakeney New Cut to Morston Creek. It matches a continuing ridge that extends on westward from Morston Creek to Stiffkey Freshes and along to Warham Marsh which faces the open sea. This ridge is created by open-sea wave action during storm-driven spring tides; that it exists from Blakeney to Morston, now sheltered from this process by Blakeney Point, shows clearly that the Point wasn't always there. Behind the row of suaeda bushes on the marsh opposite the Blakeney Hotel there are a number of wooden posts embedded in the marsh. When I was a child they stood about two feet in height; today they protrude above the marsh by less than an inch; that's how much the saltmarsh has accreted in sixty years. A seventeenth-century map of Salthouse marshes shows the shoreline considerably further north, and running through it, deep and wide, the channel from Cley. What is particularly interesting about this map is that of the numerous eyes that are shown only part of Gramborough Hill is now left. What it also shows is that most of the drainage into that old channel came in from the South, indicating that the land to the north—even at that time lost to the sea—was probably high ground, glacial eyes like Gramborough and the Blakeney Eye and which was almost certainly the source of the material that forms the shingle ridge today.

The general pattern of change on the north Norfolk coast is divided. West of Cley, the shoreline is generally accreting seawards and to the east it is currently moving inland at an average rate of about a metre a year. For the cliff line this happens in a haphazard fashion with cliff falls here and there; sometimes as much as thirty metres can fall away but then no more will fall in that location for half a century. The shingle ridge between Cley and Salthouse is similar. Although most of it moves at the same time, it doesn't move inland every year; it can go as long as twenty years with little movement before a tidal surge will roll it back twenty metres overnight. Knowing this, we are able to extrapolate backwards, so, for example, 100 years ago the Cley shingle ridge was roughly 100 metres further to the north; I say roughly because it has retreated by at least that much in my lifetime, which suggests that things may be speeding up. Over longer periods, looking back it offers a relatively accurate guide. It is this guide that provides the thread that these stories are hung on, so let's step back into history.

# 1

**1 CE**

Coming in From Sea

The date is 9 September AD 1 (*'ante diem V Idūs Septembrīs'*). We don't know that, of course, as the Romans haven't turned up with the calendars yet, but they are fast moving north; this part of the country is still in the trailing edge of the Late Iron Age. The Glaven empties into the sea over a mile north of where Cley Beach carpark is now. A wide, shallow river winds out to a flat delta similar to, but larger than the entrance to Blakeney Harbour today. Sea trade around the coast in small boats is well established and we are returning from such a passage. We are making our way in from the north-east on the first of a spring tide with a fresh, but favourable north wind in a small, heavily built craft, with a very high bow and stern,[1] under a thin leather sail and steering with a side oar. We have a crew of four and openings for oars on either side. The shoreline from the sea bears little resemblance to the coastline we know. Behind the approaching breakers a ribbon of bright sand is visible and beyond that a more distant edge of salt flats with a few low grass- and scrub-covered hills. And at the edge of sight, blue with distance, the higher tree-covered ground. No churches, no buildings, or sign of human habitation is visible.

The unmarked entrance to the river's mouth is difficult to see, let alone navigate without local knowledge, but we have a good Iron Age pilot and we will need him. This is a large estuary that runs seaward through low sand flats with a wide shallow entrance. Getting closer, a break emerges marked either side by broken water: that is our way in, an opening more than a mile across and full of sand banks. The turbulence either side of us shows how greatly it varies in

---

1. The boat I have put us in may have been sophisticated for the period, but I am sure that ships had evolved considerably in northern Europe by the late Iron Age. Julius Caesar evidences various ocean-going vessels and describes them as being rigged differently to Roman ones; Caesar remarks that some of these vessels' timbers comprised beams a foot wide fastened with iron bolts as thick as a man's thumb. The high bow and stern and heavy oak construction would have evolved to cope with the short steep seas of the North Sea.

depth. In this small, shallow craft with a fast tide under us we'll probably be fine if we stay clear of the breaking waves. Nowhere across this entrance is there any great depth but where it is a little deeper the waves steepen dramatically but don't break; it is one of these we must 'ride' to get safely in, where the waves are not breaking will be our safest way. Just like the surfers of today we have to run in on the face of the wave. Using sail and oars we get ready; a mountain rises astern, surely it will overwhelm us, but just as it looks as if it must, we are lifted, up and up, as if in the arms of a giant, and we're flying. Such exhilaration! The boat rushes along creating a wind that blurs vision. This is the fastest anyone will travel over the ground for millennia. We seem to tumble down the face of the wave, but as we fall so it rushes on beneath us buoying us up. Judging our course through blurred eyes our pilot hangs tight to the steering oar, one slip at this point and we could broach and our adventure will end.  As the wave loses its energy in foam we emerge into the wide, calm estuary, relief all round to finally be in quiet water, its colour and the patterns on its surface offer clues with which to navigate the shoals. With the wind astern and this fast tide we fly along and should be well up the estuary before it turns. The tidal Glaven holds a great deal of water, and when the ebb sets in against the wind, the slog[2] could make things difficult for our primitive little vessel.

We are in the channel proper now; beyond its wide shorelines the sand flats stretch away east and west. Off to the east, soft with distance is higher ground, but to the west the flats run to the sky. After no time at all we have covered a mile. The banks close in a little and the scene either side alters as we pass through a narrow ridge of sand, shingle and suaeda, a ridge pushed up by waves at high

---

2. Slog: Steep standing waves that are produced when the tide is running into the wind. Such conditions were used to great effect when defeating the Spanish Armada.

water stretches off into the distance either side. Once past this the sand, flats are replaced by mud, greened with eelgrass and patches of samphire. Suddenly the sky darkens and is filled with the roar of wings as huge flocks of waders take off at our passing. The tide is carrying us quickly and a major branch in the channel is coming up, one stream going off to the east and the other onward into the valley. We'll need to navigate our small craft across the tide to the west or the power of the water pouring into the east channel will sweep us off that way. Using sail and oars, our crew work hard to carry us clear.

The muds, like the sands, stretch to the horizon, just a few low grass-covered eyes in the foreground and sand dunes at the limit of visibility. To the east the wide channel snakes away between low hills that are rapidly becoming islands as the tide spills out to cover the flats around them. Grass-covered with clumps of gorse and elder and fringed at the high-tide mark with suaeda, these hills host small clusters of thatched roundhouses: these belong to families our crew know, but they are not our destination today.

Another mile and we are leaving the wide expanse of mud and sand behind and the channel narrows as we enter the river valley, its shoreline fringed with the beginnings of saltmarsh, its sea purslane lifting with the tide. Rounding the corner where Cley Mill will one day stand the ground rises, and ahead of us a landscape of trees and small fields run down to the shore on both sides. Adjusting the sail, we tip the bow toward the eastern shore, as we do so a small bight[3] becomes visible on the western bank that leads back a short distance into the surrounding fields. It's not large, but offers some shelter from the north, which

---

3. The small bight on the western shore. Today this comprises parts of Leatherpool and Chapel Lanes and runs up as far as the Glebe Barn in Wiveton.

SHIFTING SANDS

accounts for a few small fishing craft pulled up on the beach at its furthest end.

Having completed our course change we must stay this side to avoid the mud island that is just disappearing under the tide and fills the middle of this river. The fresh breeze that has helped our passage has strengthened and there is little shelter here when the tide fills this estuary. Fortunately, our destination is not far ahead. As we are approaching another larger inlet this bay cuts back into the higher ground and affords good shelter from onshore winds. Swinging in out of the fast-moving tide, our crew can use the oars now to assist as we weave our way in between small boats and draw near to a settlement of roundhouses set back from the water's edge. These vary in size but are generally a bit bigger than those we saw on the islands as we came in. A little higher up the slope one larger house dominates the settlement. A timber jetty juts out from shore and here we can tie up. We have arrived at the nascent Port of Cley and have come ashore just a short distance from where its mighty church will one day stand.

People are busy as the day is ending; animals are being brought in for the night and smoke is seeping through the thatched roofs of the roundhouses like smouldering bonfires. Some of our crew are being met by their families and go off to their small homes; as guests, we are escorted to the large, roundhouse further up the slope to meet our hosts for the evening. In the gathering night lit only by small oil lamps it seems very dark as we enter. Adjusting to the gloom we can see a fur-covered ledge around the outside and a large circular hearth with a fire in the centre; the flickering light from the flames adds to the weak illumination. The smoke makes our eyes water and the smell of animals, people and dung is quite overpowering, but I'm sure if we stayed here for any length of time we'd notice it no more than these people do. Large wooden boards of food are brought in; it is clear they intend to treat us as honoured guests.

This is very much a community and slowly the building fills with others who have come to greet us. The headman appears and the talking begins, little has altered in their world for a very long time, but they sense that change is coming. We ask how much they know about the Romans and are surprised to discover that they know them well and have frequent contact with them: a small Roman trading vessel had visited just a few weeks before. But the uneasy relationship of diplomacy and trade the Romans have cultivated with Britain for so long is deteriorating and a growing imbalance in favour of wealthy Romans is fuelling unrest; news of it regularly reaches this isolated place. Much sooner than we are used to, our hosts make their excuses and retire for the night. We are given straw-filled bags and some already inhabited furs and told we can bed down round the fire.

At first light this place is already busy. These people work to circadian rhythms, cattle and sheep are being taken out to graze and small fishing boats are leaving on the ebb. Most will fish within this great estuary and if the wind drops a few may venture out to sea. We will take the morning to explore more of this Iron-Age world on foot. Walking down to the quay, we look out across a valley no wider than today but totally strange, nothing to give us any clue as to where we are. No Wiveton or Blakeney churches and no cottages or houses, some small fields run down to the water's edge and scattered among them a few roundhouses like those at our backs. As the tide ebbs the island of mud emerges dividing the stream in two. The furthest of these is wider and will one day serve the quays at Wiveton. Further up we can see that they come together. Beyond, a smaller channel (although still surprisingly large to us time-travellers) wanders its way inland between saltmarsh up to and beyond where Bayfield Park will one day be.

Returning through the settlement we follow a rising path[4] past where Cley church will one day stand. This gives us an even better view of the estuary and the hinterland beyond. Across the water to the west, we can better see the small almost allotment like plots that run down to the waterside. To the east the ground rises to trees and heathland. Ahead, a wide sky bright with light thrown up from the sea and the vast tidal landscape that awaits

---

4.   Path: one day to become Church Lane.

us. After a half-mile or so we come to an escarpment (Hill Top, as it is known today). The view that greets us is beyond anything we could easily have imagined. The sea, which today is barely a mile off, is just a thin ribbon of blue at the limit of visibility. If we had been standing here just a few thousand years ago it would have been more than a day's journey further off.

For 100,000 years the northern lands were locked in ice. The earth has been in an ice age for millions of years; while there is ice at the poles, our world remains in an ice age. But this big freeze has forgiving moments when it eases its freezing grip; these brief intervals of warmth are called interglacials and come along every 100,000 years or so and can last for anything between 10,000 and 15,000 years before the ice returns once more. The last glaciation retreated 12,000 years ago and as it melted a new sea was created. Many ancestors of the people we are staying with lived and hunted the wide grassy plains that once stretched from here to Denmark. As our planet moves through this interglacial period, temperatures will continue to rise, more ice will melt, sea levels will get higher, and so the shoreline will continue to move landward.

We know from our exciting passage in from sea yesterday that this is a very exposed estuary, but it will not always be this way. It will remain open for several centuries yet (as we will discover) and will remain exposed and shallow with passage from seaward always hazardous. But eventually the sheltering arm of the Point will extend to protect it, and the Glaven estuary will become Blakeney Haven. Much will change during the centuries we are yet to visit. In front of us and about a mile off, is the wide channel, that we gave a wide berth yesterday; it is deep and wide and bears much of the tidal prism[5] for this huge area. One day, 1,000 years hence, it will carry ships up to Salthouse and Kelling. Scattered among the mud and sand flats are the small grass and scrub covered islands we saw yesterday. These are glacial deposits, pockets of sand and gravel, that were caught up in the ice sheet and then dropped like hot potatoes when it melted. As the shoreline retreats landwards so these will begin to have a major effect on how the coast evolves. When the sand flats eventually disappear and the sea reaches the first of these hills they will offer more resistance, the shingle and stone they are made of will form a protecting ridge and how the coast evolves will begin to change. It will still move landward, but more slowly, and some of the eroded material will be carried westward: longshore drift will begin and the first nub of Blakeney Point will begin to form.

---

5. Tidal prism. The volume of water held in an estuary at high water.

# 2

1250

Delivering Around the Haven

We have pressed all the right buttons and our time machine has carried us forward almost thirteen centuries: much has changed. The glacial moraines we saw to the north of Cley and Salthouse have been eroded as the shoreline has moved inland, providing the offshore deposits that, shaped by the wind and sea, have grown into a shingle and sand-dune spit that has extended a protective arm to embrace the Glaven Valley. It stretches from the same point as today (Kelling) at the eastern limit of the haven to opposite the newly built Carmelite Friary.[1] The curling end of the spit (Watch House Ridge today) forms the entrance to Blakeney Haven and lies to the north of the small agricultural settlement of Snyterley.[2] Although the shoreline has

---

1. Friary Farm today.
2. Originally Esnuterle then Snyterley and subsequently Blakeney.

retreated since our last visit it is still a good half a mile further to the north than it is now, creating space for a large body of deep water that stretches from the Haven entrance to Salthouse. Here ships can anchor in safety and float at low tide, making it one of the largest areas of sheltered tidal water on the east coast of England. Wiveton and Cley have been transformed from tiny settlements into flourishing ports. Salthouse too is a port, but not on the same scale. The wide channel we had to avoid as we came in from sea on our last visit flows eastward from the Haven passing some distance to the north of Salthouse. A smaller channel reaching southward runs closer to the village. Because of the glacial tills that lay to the north, this part of the Haven would have afforded a sheltered anchorage for some centuries before either of the other towns rose to prominence and would undoubtedly have been used by both Vikings and Romans.

Gyles Dobbe of Salthouse is a farmer, butcher and salter of meat and fish. Every week, weather permitting, he delivers these items to the ships and settlements of the Haven. His small double-ended boat has a sail but is not very good to wind'ard, so oars and a long pole assist. We will accompany him on his weekly journey.

Once the boat is loaded we wait; in this world, wind and tide set the rhythm that all must step to. This south-west wind will render our simple lugsail almost useless so we'll need the outgoing tide to carry us on our way. This is a small creek compared to the main channel, but large enough for small vessels and the flat-bottomed boats that ferry cargo in from the larger ships at the outer quay. The pilot of the one we meet on our way down waves and wishes Gyles a safe passage; he is making great progress because the wind is in his favour. Over our shoulder as we swing out into the main channel a couple of the large seagoers are being unloaded and another is being swung, ready to leave on this ebb. Once into the main channel the current takes a firmer grip on this heavy boat and Gyles needs only to use the oars to keep us on course. This wide, easy stretch gives us a moment to look at our surroundings. Ahead, a wide passage meanders west towards the Haven and beyond the first bend a distant ribbon of masts tell of the many vessels moored there. Just a little to the right of them a large area of sand dunes forms the end of the spit and hosts the settlement of Blackeneye, which is our final destination today. Stretching back toward us, the dunes diminish to become a shingle ridge that from this viewpoint disappears behind low hills of waving grass and bramble lying to our right. Grazed by sheep and goats, they are Little

Eye, Great Eye, Gramborough Hill and Flat Eye. These and the gaps between them where the shingle ridge is visible form our northern horizon. Astern of us, the channel continues eastward through saltmarsh to disappear behind a headland as it winds its way inland to Kelling. To the south over the marsh and mud, the ribbon of low, thatched dwellings that make up our village. Above them on higher ground, a lattice of wooden scaffolding surrounds the growing tower of our new church. But the tide is carrying us fast and we must stay sharp.

Two miles are covered in a very short time and we are quickly out of the Salthouse Channel and into the vast expanse of water that is this great haven. The bow lifts, the rise and fall as great as if heading out to sea. Gyles hoists the sail to make the boat's movement more comfortable. We have room to make more use of it now, but this will be a more lively journey until we reach Blackeneye more than two miles away.

We are fortunate that the breeze is not strong, because this is a large body of water and with the wind against the tide it could become as rough as the sea, but we'll be fine; in any case at this point we'll gain some additional shelter from a group of islands flanking the Cley channel, the larger, Thornham's Eye,[3] which will become strategically important in the years to come, a chapel will even be built on it to bless those braving the perils of the sea. As they slip away astern we enter the main body of the Haven, ahead a number of cogs[4] and smaller fishing doggers are lying at anchor. Two of these are expecting us before they up anchor to leave on this tide. We are bearing down

---

3. The high ground that now sits to the north of the new Glaven diversion and is the site of the old chapel.
4. Cog. Most ships of this period were of a design called a cog. These were not big Tudor galleons but small, clinker-built vessels, with a square sail on a single mast, partly flat-bottomed and with a primitive stern castle. Unlike their predecessors (the Viking ships) they had straight stem and sternposts. They could fit into any yacht marina today.

fast on the first one and getting alongside will require the use of the sail; on approach Gyles turns back into the current and hoists it, this aided by the oars, will give us enough way[5] to stem the flow and come gently alongside. The master of this ship is a very grumpy William de Orfere from Berwick, who only recently was the victim of pirates just a little way up the coast where he was robbed of 320 marks worth of cargo. He knows that those who robbed him probably hide out on Blanc's Isle. So a fast and silent transaction takes place, he ought to know that Gyles would not have been involved

---

5. Way: momentum.

in such activities but in this world trust is in short supply. Our next customer, John Curlew, is a local man who trades between the Haven and ports to the south, his cog lies further downstream, so without further ado we cast off. The current soon takes us down and we repeat the same manoeuvre to come alongside. Supplies are loaded swiftly as John also wishes to get underway as soon as possible. This done, we cast off and head toward the end of the spit and Blackeneye, or Blakeney as its increasingly called. On a map in Salthouse old church, it was called Blaca's Isle. Gyles' brother lives there and is also a merchant. The tide is flowing faster now and running against the wind it is getting rough; we may be in the Haven but it's like being at sea. Gyles shortens the sail, which gives a bit more stability, but this little boat is still butting its way through some steep seas and throwing spray over us as we plunge into the waves. Changing course toward our destination, the boat heels over and the wind'ard side lifts, offering a little more shelter as the spray flies over our heads. The dunes ahead grow closer and soon we are in the shelter of the small creek that leads through saltmarsh and up to the settlement.

It lies within the curve of the spit, scattered among the sand dunes and along the sides of the creek. As we come in it is a hive of activity, a whaler in need of repair is settling onto large logs as the tide retreats, fishermen are returning from many days at sea, carts and wagons line up to take their catch back along the spit or to the merchants that supply the ships. This place is the hub of the Haven, as big as Salthouse or Snyterley, only Cley and Wiveton surpass it. All the livings connected with the sea can be found here; fishermen, shipwrights, sail and net makers and, as William de Orfere knows, more than a few miscreants whose activities benefit from a little separation from the mainland. We secure the boat and head to the home of Gyles' brother James. This is not a flint and pantile cottage but a wattle and daub structure with a reed-thatched roof. While James makes us comfortable Gyles goes to the small timber chapel to give thanks for a successful day's trading and a trouble-free passage. Not something that can be taken for granted in these lawless times. He will also pray for fine weather for the morrow when we must set off on the flood to make our way up to Snyterley. When he returns a friar comes with him.

Leaving Gyles to talk business with his brother we take up the friar's offer of a walk through the sand dunes that cover this end of the spit. At the very end we climb to the top of the highest dune. Lit by a westering sun this vast landscape glitters, a seemingly endless expanse of sand and water throwing light into to the sky and where it touches the horizon a thin indigo line that may be dunes, though at such a distance there is no way to be sure. In the foreground the wide channel and entrance to the Haven. A solitary cog sits on the sand ready to take full advantage of the tide to head south tomorrow. To the east, an area of saltmarsh fringes the spit giving way to mud banks as it forms the northern shoreline of this broad, low water inlet.

Since we arrived other vessels have dropped down on the ebb from Cley and Wiveton to join those we passed when coming here, all poised to depart on tomorrow's tide. Further off we can see Salthouse in the distance and a bit further to the right, the entrance to Cley Channel. Coming westward the ground rises slowly to host the wall of the Carmelite Friary and its buildings, and behind them on the highest point the church of Snyterley. Not the magnificent structure that graces this hill today, but a much smaller affair built in the mid-1200s on the site of an older church. The wall of 'The Friars' leads the eye down to Snyterley itself, a modest village very much like Salthouse.

Great changes have occurred since our last visit so many centuries ago. It took a thousand years for the sea to reach those glacial hills we saw coming in from sea, but once it did the evolution of this stretch of coast took a dramatic change. As the shoreline moved inland, material from those hills was pushed into a ridge and a spit began to grow. The entrance to the Glaven River was slowly moved westwards and the ridge protected the river from the open sea. The friar tells us that at the same time as all this was happening the foundations of Norwich Cathedral were being laid. Salthouse, he says, had been a sheltered inlet long before the major ports of Cley and Wiveton were born: both Romans and Vikings took advantage of its shelter. But as the spit continued to move west, so the area of sheltered water behind it grew larger and larger. Over the next three hundred years it broadened to great size and both of these ports flourished, making Blakeney Haven as important as any in the country.

'Why the name Blakeney Haven', we ask, 'when the village of Blakeney that we know is so far in the future?'

The friar explains: 'As the map in Salthouse old church tells, the spit and its sand dunes were known by different names to different people. Some of those who visited called it Blaca's Isle and some Blanca's Isle. Blaca could refer to the black mud flats of the Haven, Blanca to the white sand that forms the dunes. In my time I have heard it called both.' To which one of us adds, 'much as people query Cley, Cly or Clay, in our time.'

He continued. 'What is more certain is that as time went by that name changed to Blakeneye, then to Blakeney, which has become the name for the Haven. Not for many years would the name attach itself to the village we know as Blakeney.'

'These sand dunes are like the foothills of a mountain range', we say. 'How did the spit get so big?'

The Friar expounds. 'This whole place is made by the sea and the wind. I'm old now but I have seen what happens', he says. 'The shore changes very fast, the sea waves and tides bring in offshore shingles and gravels that get spread along the shore and form the backbone of the spit. The dunes are med by wind and tide and are added to the head of the spit by all that sand out there to the west. This here entrance is a daleth,[6]

---

6. Daleth: delta.

wide with sand shallows all across it, but it's always on the move, changin' over time. When the tide's a-flowin' out from the Haven the flow at sea is always running down that way, to the east, this pushes on the water what's a-comin' out a' the Haven sideways and so the channel gets pushed to the east: it takes years to do it. It only moves tens of yards in any year, but after half a hunard years or so it ends up winding a tight and winding course around the head of the spit. Then at a chance time, a large wind-pushed tide: too big, too strong and anxious to get out, that it won't follow such a winding channel and will force its way back straight out to the northwest and that that'll all begin again. Once that's moved, all that lot of sand left on the east side of the new channel is stuck onto the head of the spit and then as the years go by that gets blown into these here dunes. I've seen the channel move twice in my lifetime. There's a mighty lot a sand out there to the west and when it's small tides, an' the wind's a-blowin' hard across it, that'll pick up a blizzard a sand, sometimes that kin blow for days, then that'll move mountains of it. A fisherman left his boat on the side of a creek once and that were never seen again completely buried that was.'

It is almost dark as we return to the settlement. James's wife has prepared supper, and all gather around the simple table. Although James was born in Salthouse he has lived much of his life here on the Black Isle. It is a place unlike any other; separate from the mainland in many ways other than by a tidal estuary. Gyles asks him to tell us a little of what life is like on this lawless finger of sand and shingle.

James looks up and touching the side of his nose says, 'if you want to stay a liven' hereabout you gotta keep "this" outa everything. It's a crocka vipers and villains that'll slit your nekke as soon as look at you. There's a truce here twix those who wrek and raid and those what don't: but it's hard to tell t'other from which. The sole intent of the scum that use this place is piracy and wrekin'. Thas bin goin on a hundard years amore. When there's a hard north blow, foreigners run in here for shelter; cause there's none to be had anywhere on this coast for miles. Then they'll set on 'em; kill em, take the cargo and break up the ship, I reckon there's many a good ship gone to the deep for fear a-comin' in here. This might be called a Haven but thet ent no such thing.'

We say that there seem to be plenty of other less barbarous things happening here.

'Oh yes, acorse there's other work to be had. There's hundards a fishin boats: some local, some furriners, some take their catch up to Snitterley an Cly, others bring it here to sell to the ships. They dry the nets on the marsh or on them pusts out there. There's others wat make an mend ships, mostly walers[7] they make them course they can soon be med warlike when the King's enemies are a-comin. But wrekin an skekking[8] is why most on em are here. Awhile ago they took a ship right by here, kilt all the crew and took her orf to Aberdeen, sold the sailors clothes, then when across to Norway and sold the ship. Later another ship loaded with corn, wine and other stuff got inta trouble in a storm, they stripped her of all her cargo and then broke up the ship. A year later two German merchants were attacked by forty men, just outside the harbour. Another time they took two ships together, fifty men overpowered the crew and brought them into the Haven and there robbed them and carried off the spoil. There's been countless ships taken along this coast, and in the Haven itself; they run em a ground a purpose then strip em clean. Murderers all! But, we say, 'there's a small chapel here and a priest; surely that's a bit odd in such a place?'

---

**7.** Whalers: these were called Balingers.
**8.** Skekking: raiding, attacking.

SHIFTING SANDS

'Nay not at all', James replies. 'There's another on Wiveton Bridge, the priests are paid to bless 'em and pray for 'em, for safe passage an all that. Don't do 'em much good that I can see!'

James talks on into the night, telling of many raids and ships wrecked and robbed, this is such a different world. It's been a long day, and another awaits tomorrow so we ask Gyles if we might make our apologies, and we retire for the night.

# 3

Concluding the Deliveries

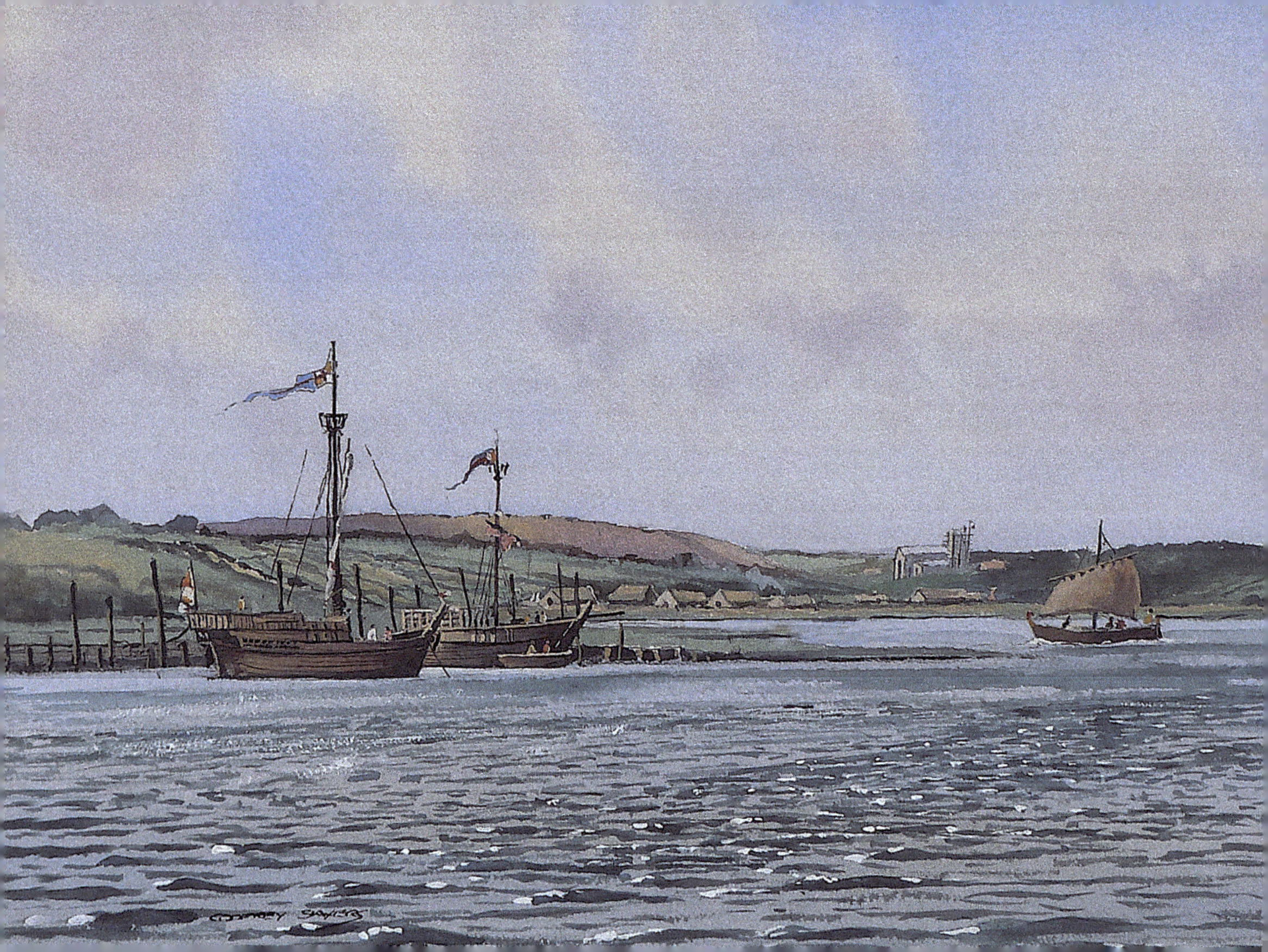

After a late night it's an early start, a hasty breakfast then quickly down to the boat. Gyles and his brother are there before us. James has goods of his own to go to Snyterley and these are being loaded. The wind, a moderate *sou'-sou'-west* at the moment, will help us against the incoming tide, but it can often increase during the day, so James will join us on our trip up to Snyterley. His help will be needed for the hard rowing we'll have, getting us across the tide and over to the small creek (the Benhaughe Stream) that leads up to the village. The friar who ministers the settlement's small chapel comes rustling along. He's coming with us; his ministry has come to an end and he is to be replaced. At least a dozen doggers[1] accompany us down the creek

---

1. Doggers: small fishing boats.

to join many more coming through the haven from the other villages. Some will be out for the day; the larger ones go much further and could be away for a week or more. Even with some help from the wind, progress down to the Haven is slow as the tide is running in fast; James assists by pushing with the pole. Out in the Haven, ships are making ready to leave as soon the ebb begins. The cog we saw sitting on the sands by the estuary mouth last evening will be long gone south on the flood. As we swing out into the wide body of the Haven, the flow catching us, our course now north-west, means that the sail can give a little assistance. Gyles quickly raises it as James picks up the long pole and using it keeps us close inshore to cheat the flow.

In a bit we'll be heading almost due west so the sail will not be of much use. In front of us a headland juts out into the stream; the tide pouring around this is creating a powerful back eddy and James means to take advantage of it. Its effect is soon felt as almost with a jerk we suddenly gather speed. As it carries us toward the tip of the spit Gyles asks the friar to take the tiller and if any of us can row. We all can but only two are needed. He positions us ahead of the mast and then sits with James astern of it. He tells us to keep our strokes in time with theirs and after a few minutes of practice we are out in the fast-running stream. This is a loaded boat and it catches the current that wants to sweep us off into the haven. The breadth of the channel has to be crossed before there'll be any easing of the tide. It is strenuous work and seems to take forever, so it's a great relief to finally get out of the turbulence, but despite some very hard rowing we've been carried some distance down stream. Once more keeping out of the flow we have a hard row back west. The friar is steering us towards the entrance of the Benhaughe Stream, which will carry us up to Snyterley. Once the tide pouring into this narrow creek catches hold of us the oars are shipped and Gyles hoists the sail again. The little boat, suddenly picking up speed, reminds us once more of how all that happens in this haven is entirely at the mercy of wind and tide.

The creek is no wider than it is today and meanders its way first through sand flats. We fall in line behind two other traders on the west side to keep clear of the weaving convoy of small doggers sailing out to join the many others we have seen leaving for the day. After a mile or so Benkallpoynt[2] is reached, now the

---

2. Benkallpoynt would have been just a little north of the end of the New Cut today.

channel turns south and the sand flats give way to saltmarsh. One of us comments on how quickly the marsh[3] is disappearing under the rising water. James says that when he was young there was no marsh, it was all just mud flats.

'Each tide', he says 'makes it grow higher.' He added, 'There's been a lot a concern that this creek was silting up, so a few years ago a long circle of bank was built (Pilot's Path today) right out across the marsh to connect with the natural meols[4] that run as far as the crick at Morston to make sure that all the tide that covered the marsh ran out the same way as it came in and that's made a great difference.'

The village soon appears straight ahead, a cluster of single-storey thatched dwellings clustered either side of a track, that comes over the hill to Snyterley from Wiveton. Past the small village church down the slope through the village and through the arched gateway we can see in the wall (that is still under

---

3. 'Saltmarsh' grows by a process called accretion. 'Every tide that covers it carries a burden of fine silt held in suspension, when it spreads out covering the marsh, the current is stilled and the sediment drops. This process can raise the height of the marsh by as much as two feet in a century. So the marshes we are travelling through are many feet lower than they are in our time.

4. Meol: the remains of a natural ridge pushed up by the sea before the Point evolved to protect it.

construction) extends a protective arm from the Friary to the east. There are no proper quays yet but there are numerous short jetties along the ribbon of high ground[5] that extends from the track that comes from the gateway in the wall.

We are expected. The merchants that Giles supplies post a boy on lookout for his sail, so along with the friar's replacement they are already on the Carnser waiting. James swings the boat neatly alongside Gyles' jetty. The salt meat and fish we have brought are loaded onto barrows and carts, and supplies for James' business are loaded aboard. Several locals who know Gyles and James come over to catch up with news from Blakeney, including Mathew Curlew, a friend of James's, whom he greets with a familiar bit of banter.

'Ahoy there Mathy boy, hev you paid any a that money back yet?', referring to a longstanding debt, owed to men of Blakeney by the men of Snyterley. Many years before, two men from

---

5. This ribbon of higher ground eventually became known as the Carnser, more recently it has been referred to as the First Mariner's Road and is still the responsibility of Norfolk County Council Highways department. Over the years as the area to the east became reclaimed from saltmarsh, so the name has expanded to cover the whole area from the channel to the sea bank. The Carnser carpark.

Blakeney had paid substantial dues to the King on behalf of men from Snyterley and have never been paid back.

A keen eye is kept on the tide as the ebb is needed to carry us back, but for the moment the flood is still running, so there's time for a jug of ale and a chat with old friends, who have much to tell. Some days before, a band of brigands landed to the west, and joining with some local ruffians had been pillaging villages all along the coast. Snyterley has escaped so far but its inhabitants are fearful that such an attack might be imminent. Blakeney, perhaps because it is almost an island, may escape such an attack and with so many brigands living there, they might not be such a push-over anyway.

We remark at how busy the quayside is, small ships being loaded, the last few doggers pushing off and a bustling air of purpose about the place. Mathew grunts and begins a sorry tale.

'It may seem busy now', he says, 'but this place is just a shadow it what it once was. A while ago a hundred men from here were a fishing by Doggerland when a fleet of Scots enemies set on 'em. They fled to the port of Wynforde in Norway hoping for sanctuary there. But that weren't God's will that day; 500 armed men of the Hanse that come from Bergen, came on them and tied their legs and hands and threw 'em into the sea before the port and so they were murdered every one on 'em. And ever since they of the Hanse have been bold to harm the English seafarer.'

This place will never be the same again.

The tide has turned, the news is caught up with and the trading done. James bundles the friar aboard and pushes us out into the channel. The sail fills, the tide takes hold and we're off, flying away down the creek toward the Haven.

Now both wind and current are with us, so the distance to the Haven will soon be covered. This quiet easy stretch with no rowing needed allows us to fall into conversation with the friar whom James has introduced as Sebastian. Our first question is how he sees his role in this Godless place. His eyes widen:

'Godless?' he exclaims, 'this place en't Godless, excepting those that belong to the devil a'course; seafaring men are God-fearing men. They may go out and take ships, but them ships belong to the King's enemies, they do make a mistake from time to time as not all on 'em fly the right colours but in serving the King this way they are doing the Lord's work, and getting revenge for the terrible things that have been done to our people.'

Clearly this friar sees no conflict between robbery and slaughter and the will of God, but we have military chaplains that go to war with our servicemen to comfort them if they feel guilty for taking a life, so perhaps not a lot has changed. It is easy to kill and abuse if one can demonise the victim.

We come out of the Benhuagh Stream like a cork out of a bottle, but immediately hit the powerful ebb running out of the Haven. To the north the last of the cogs are leaving, an armada of square, bright sails vivid against the sand and sky. This time our sail will help us win against the tide, and after a half-hour or so making our way east, we are ready to head north across the Haven to Blakeney.

After a second night spent with James and his wife we will continue on with Gyles as he makes his deliveries around the Haven. His next stop is Wiveton.

The day dawns bright, the light breeze has veered to the north from yesterday and the tide is pouring in. We wave farewell to James as we leave the mouth of the creek and once more, entirely at the whim of the wind and a distant moon, our little boat is carried on its way; all Gyles has to do is make sure he keeps clear of the ships that remain at anchor. A moment to relax and be carried by the flow across this amazing inland sea. The Haven will never again be as big as it is now; we might be setting off across The Wash, so vast is the space before us. The high ground behind Salthouse is soft with distance and the spread of moored vessels stand dark against the glitter of the bright morning sun. On our left, sharp and clear are the dunes of the Black Isle, and far off to the south the high ground east of Snyterley.

It is difficult to judge speed in such a wide expanse, but we are moving over the ground fast, but even at such a good pace it is half an hour or more before we reach the point where will turn off into the Cley Channel. Fortunately, the wind has continued veering north, so we will still feel its benefit after the course change. We have kept an easterly track through the Haven and are approaching the small islands we saw outward bound. After passing the first one, Thornham's Eye, we turn south, following the deeper water that leads into the valley and on to Cley and Wiveton. After a mile or so as we leave open water, another course change is needed.

On our last visit we turned east and followed the Holfleet to visit the small settlement that has now become the port of Cley. To get up to Wiveton Quays we must keep to the west and enter the Milsteade.[6] This done, we soon pass the first of Wiveton's buildings,[7] a large flint and limestone structure situated at the entrance to this channel, originally built as a fortification against raiders, it is now a merchant's house. From it, a substantial wall skirts the waters'

edge for a distance up to the sheltered inlet[8] we saw on our last visit, where smaller fishing boats are built and repaired. A number of other thatched buildings are visible behind the wall.

Over the water, the port of Cley is strung out along the shoreline, dozens of small thatched buildings behind the many masts of cogs and doggers that sit at its quays. As we move along the new church, surrounded by timber scaffolding comes into view tucked into the

---

6. The Milstead: this is the established name by 1586, so either this or some precursory form would have been its name at this time.
7. Situated at the bottom of Cley Hill on the corner of Leatherpool Lane and the A149 Coast road.
8. The small bight on the western shore. Today this comprises parts of Leatherpool and Chapel Lanes and runs up as far as the Glebe Barn in Wiveton.

bay. The tides are large at the moment so there may be other bigger vessels tied up at Wiveton Quays. We will bring our boat ashore a little further up, just short of where the track comes down to the stone bridge. As we round the bend there are indeed some cogs and other boats at the quay, but the waterway is wide here so there is plenty of room for us to pass and swing into the small bay by the bridge. The last wagon to get across the causeway[9] from Cley before it went underwater is making its way up the slope ahead of us. Wiveton is the largest, busiest and noisiest place we have encountered so far. A fair is taking place in a wide space in front of the church; the smell of animals and cooked food greets us as soon as we pulled ashore. Securing the boat, we go up and cresting the slope enter a world of colour and movement, tents and covered stalls surround livestock pens and crowds are milling among

---

9. Current belief is that the second bridge (the one gives its name to Bridgefoot Lane) was made of wood. I have some reservations about this. About forty years ago, along with a couple of friends I excavated around where this bridge would have been and we found some quite substantial flint work. John Wallace always claimed that 'Wiveton Stone Bridge' was referred to in this way to differentiate it from the other, wooden bridge. However, between the two bridges was a causeway of stone: almost certainly brought up by boat from the beach. This might also have been referred to as the bridge of stone, or the stone bridge. On the 1586 map of the Haven it is shown pretty much the same as the existing bridge (see map). We will probably never know, but for the purposes of this tale it is stone.

them. Squealing pigs, bleating sheep and the barking of dogs sound out above the babble of human voices.

Gyles goes off to meet his customers, some of whom live in the village, others have come in from surrounding settlements. This gives us a chance to explore this chaotic and fascinating place. Wandering among a warren of stalls we find all manner of things for sale, clothes and spices, rugs and ropes, food hot and cold; baking bread, roasting meat, sweet meats and sausages.  And to follow, pasties, puddings, pancakes, and pies; not so different from the high street today. Tempting as all these smells are we must remember that these people have far more robust digestive systems than we do, so perhaps just a small pie. The crowd seems very mixed, but the merchants' dealers and gentry stand out, their finer clothes an instant giveaway and much that is on offer seems to be for them; out of the poorer peoples' reach.  But they, of course, can enjoy the entertainment, which is free and set to delight everyone, rich and poor, young and old: jugglers, mummers, magicians, and a dancing bear.  We are told that sometimes, filling in any religious cracks, there are mystery plays, performed by strolling players who tell stories from the Bible. They are not here today.

Gyles had hoped that the boat would be unloaded soon enough to have had the last of the flood to carry us on our final leg to Cley. But best laid plans? There may be a delay. A friend of one of his regular customers has a pig that he has sold to a farmer in Cley and would like Gyles to take it. An argument goes around the fact that the farmer could take it across the causeway himself in a couple of hours when the water drops. But he says that he has to drive some sheep he bought back to Langham and so won't have the time. Reluctantly Gyles agrees and the boat is dropped down to a small jetty where it will be easier to get the animal aboard. Even then a large and reluctant sow makes it a difficult operation. Once this is accomplished we lower the mast, push off into the tide and pole our way along to go under the Stone Bridge. It has two arches. There is very little head room under either of them; we take the left arch. We have a choice of two routes to Cley from Wiveton. But for the pig delay we might have dropped back downstream and taken a shallow passage through the central mud banks into the Holfleete, but with the tide now ebbing this will be too shallow so we will take the deeper channel under the bridge. This brings us around a curve in the river that crosses to the east side of the estuary where we come to another bridge similar to the one we've just passed under, but this has just one arch.[10] The ebb is with us now so with just the occasional push of an oar we'll drift our way gently down the Holfleet to the Port of Cley.

It's a short crossing to end a long day and the water's getting away. We tie up and are relieved to find it is much easier getting the pig off than getting it onboard. We tie it to a quayside post for the owner to collect, then the remainder of Gyles' goods for Cley are unloaded. Arrangements are made for staying the night and the boat is dropped a short distance

---

**10.** Newgate Green today.

downstream to the limekilns. Tomorrow, a shipment of lime[11] is to be loaded. The construction of Salthouse church has been a real boon for Gyles as it has meant a regular return cargo from the Cley kilns for many months.

We retire to our lodgings for another early night; it'll be another prompt start for this last leg of our journey. The end of the ebb will be needed take us down to the Haven and the boat will have to be loaded quickly to avoid missing it. The sun is barely up as we make our way along

---

**11.** Lime is made by burning marl/chalk and was used in building mortars **and as a stabilizer in mud renders and floors.**

the track beside the quay that is already busy: we are not the only ones wanting to catch the tide. An old sail is spread out in the hold of the boat and many hands make light work of getting a good load of lime aboard. Once loaded, Gyles lets go and the ebb takes us. The wind is still from the north so even with the current under us the pole is needed to move us along.  Our arrival in the lower Haven needs to coincide with the turn of the tide so the flood can be used to carry us back through the Salthouse channel. But today the north wind has produced an early flood and the choppy water makes it hard work getting across to it. The consolation, of course, is that we'll have a fast flow to take us home.

Clues that suggest that Blakeney was separate from Snyterley.

An astrolabe now in the British Museum is dated 1342 and is inscribed 'Blakene me fecit' – Blakeney made me. It could, of course, have been made by one of several people surnamed Blakeney, but it could also have been a product of a metal worker living on the spit. A mandate of oaths taken 1436 by several foreigners who had come to reside in the district showed that, two had settled in Cley, two in Blakeney, and two in Snyterley. Why the distinction between Blakeney and Snyterley?

# 4

1586

Flemish Visitors

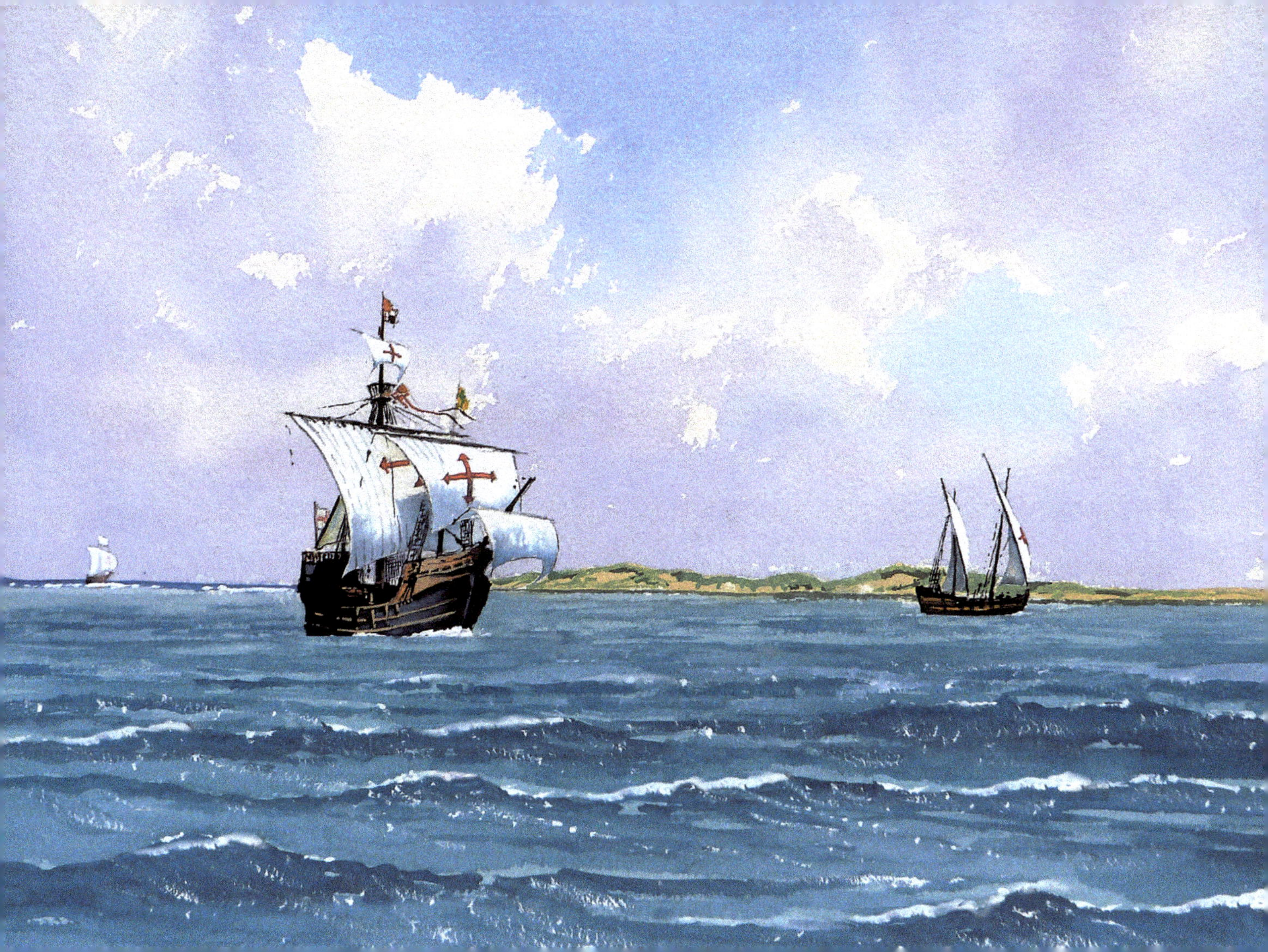

BLAKENEY ALIAS SNYTERLEY        SALTHOUSE
                    WIVETON  CLEY

In the time it took to leave the last chapter, our time machine has brought us to the later part of the sixteenth century where many more pages of history have turned. The shape-shifting of the spit has continued apace and the shape of the Haven is much changed. In geological time this a blink of an eye but in this fluid, action-packed environment a transformation has taken place. The accretion of sand waves has moved the entrance of the Haven westwards to between the Hood[1] and

---

1. The remains of the sand dunes that formed over the lateral arms that swept round to form that entrance to the Haven, remain today as the Hood. When I was a young mussel fisherman the finger of shingle that formed the actual Haven entrance then was visible above the spartina grass, and stretched from the dunes of the Hood 300 metres to the south-west forming a curling finger of shingle called Wallsback. This has all but disappeared under accreting mud. I think it still has beacon on it.

the Long Hills (at that time know as the East and West Hurdes). Three centuries of wave action have rolled the shoreline inland by more than a quarter mile, but as the spit extends west so this process is slowing. The huge area of sheltered water that formed the Haven anchorage is greatly diminished, and along with it the headland settlement of Blakeney. As the area shrank and the ship owners found it more difficult to use the shrinking Haven,[2] a slow migration from Blakeney to Snyterley took place and with the people went the name.

---

2. The term 'Haven' was, I believe, first used to describe the large inlet where ships moored, but as that disappeared it became the name for the entire area sheltered by the spit.

Snyterley became Blakeney, alias Snyterley, and then eventually just Blakeney. The demography of the Haven is changing quickly. A reduction of the tidal prism has put a brake on the maritime trade of Salthouse, brought about by a practice that would eventually lead to the demise of the entire Haven: embanking.[3] Cley had become the primary port in the Haven with Wiveton close behind. Blakeney lags a little because of its narrower channel.

---

3. The first embankment was begun at Salthouse by Sir John Heydon, in 1522. Similar operations were carried out at the same time near Cley and seem to have altered the course of the channel there. This process was seen as winning land from the sea, and for the next few centuries became a battle between maritime interests and those of powerful landowners. Our relatively tiny harbour today shows who won those battles.

Once more we are coming in from seaward, this time on a fine high summer day with a light north-westerly breeze on a spring tide and in a very different vessel. Instead of a small craft with a single sail, we have the 'Dolphin', a fully rigged Tudor trading ship. All the crew are on deck as we will be making a number of tight course changes to make passage up to Wiveton. As we run in, our first sight is of a harbour entrance very much like the one we know, sand dunes on our left and a wide expanse of sand on the right. But these sands still extend as far as the eye can see, and the channel we enter is wider than the Thames. Ahead, a smaller ship, pennons streaming, tips into the wind as it turns to enter the Benhaughe Stream to Blakeney. In front of it is a vessel smaller than our own, but the small flag she has flying off her mizzen tells us that she is bound for Cley. This pleases the captain as it means he won't be competing for turning space when we arrive at Wiveton Quays.

In the distance the sun is catching the thatch and terracotta roofs of the cottages in Blakeney. The dunes quickly drop astern and are replaced by an open spread of sand and mud flats (the East and West Meales). To the west, a large sweep of sand, to the east mud flats and in the far distance the high ground of Salthouse.

The crew shortens sail quickly as we approach Eastgate Point making a change of course to navigate a bend in the channel past Catsarse Beacon and onto a more north-easterly heading. We have barely completed our turn when the smaller Benhaughe Stream slips by and the small ship we noticed earlier is making fast passage through it to Blakeney.[4] Our ship is of more than 100 tons, more than can be easily accommodated at the quays there which are currently in poor repair[5] and only able to take smaller vessels: so our destination on this visit is Wiveton. As we come to the end of this broad reach the crew are busy adjusting sails as we are fast coming to the meeting of two channels. If we were to continue on our present course we would follow the channel to Salthouse. Here we are to pick up a pilot so a gentle turn to let the wind out of our sails slows us enough for him to get aboard. We now head south around Thornham's Eye, only a few metres above the high-water mark it is made conspicuous by the small chapel[6] crowning its highest point. To the

---

4. Although Blakeney was becoming a more significant port, the channel to its quays was much narrower than that which served Wiveton and Cley.
5. The quays were repaired a few years later and possibly the first dredging of the channel took place at the same time.
6. Since the recent diversion of the Glaven it is now situated to the north of the Cley Channel. It is believed to have had a companion chapel on the other side of the old channel on Cley Eye.

south a mile ahead is the spread of the valley. It looks wide and deep, but we must navigate with care as the Cley Channel divides into two streams, each of these we have used on our previous visits. Our pilot must beware of the shallow water that lies between. In days gone by, if a pilot put a ship aground he could be beheaded; hopefully that won't happen today.

The Holfleet we know would take us to Cley, but we must get over to the Milsteade, the larger of the two channels, if we want to get to Wiveton. With the wind astern the Dolphin is moving over the ground too fast, so sails must be shortened to slow us down.

The town of Cley is strung out along the east side of this wide expanse, its cottages following the course of the channel, the roofs a soft mixture of terracotta and thatch. Many vessels of all kinds line its wharves: fishing boats, local delivery boats, traders large and small are all tied up. And behind, showing above a rise in the ground, the tower of its grand church. This is the largest town of the Haven and since the decline of Salthouse has been its principal port, but now, due to shifting sands, Wiveton has become the busier port, berthing bigger ships and more of them.

Normally we would have dropped anchor before the streams divide, but this tide will

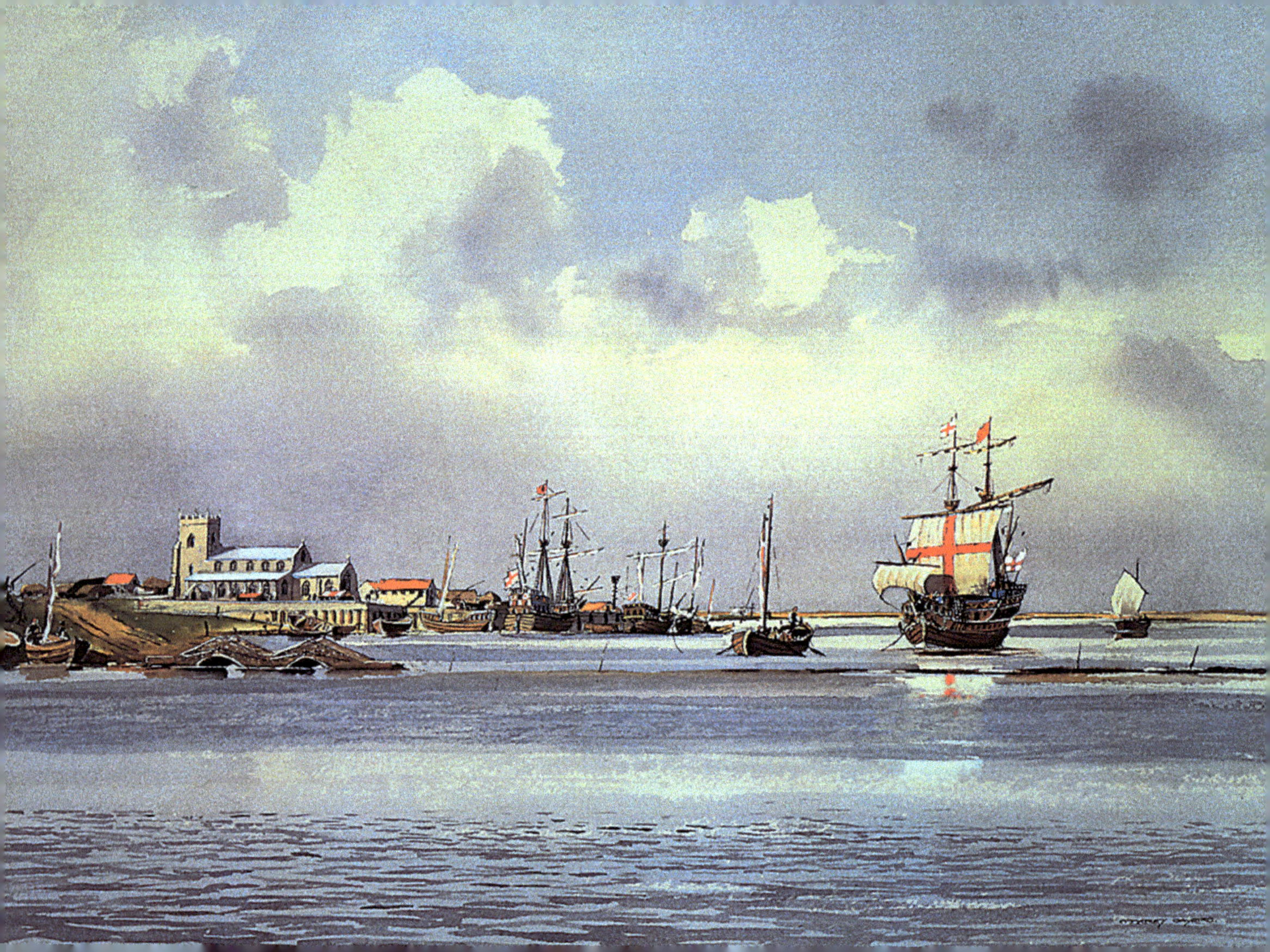

allow us to reach the quays at Wiveton. With our reduced sail we ease across to the west to enter the Milsteade. On our right the remnants of the once-prominent thirteenth-century building remain, much of its stone taken for other uses in the town.

As we gently make our way in, so the inlet we have seen before comes into view, still used as a small sheltered anchorage. We can now see the masts of ships at the quays ahead and Wiveton church standing grandly at their backs. Fortunately, there are only a couple of large vessels tied up at the quay and several smaller fishing boats have been moved to give us room to berth. We are met by a small but heavily built craft with four rowers. A rope is thrown to them and they help in bringing us toward the quay. With the wind astern we use the mizzen sail to bring the stern around. As we drift closer to the quay, ropes are thrown to those ashore. As these are caught we are slowly pulled in till we come alongside.

As well as cargo we have two Flemish merchants aboard, Henryk Peeters and Simon Martens; both are members of the Hansa.[7] Henryk, has regular business with the Glaven

---

7. Hansa: the Hanseatic League.

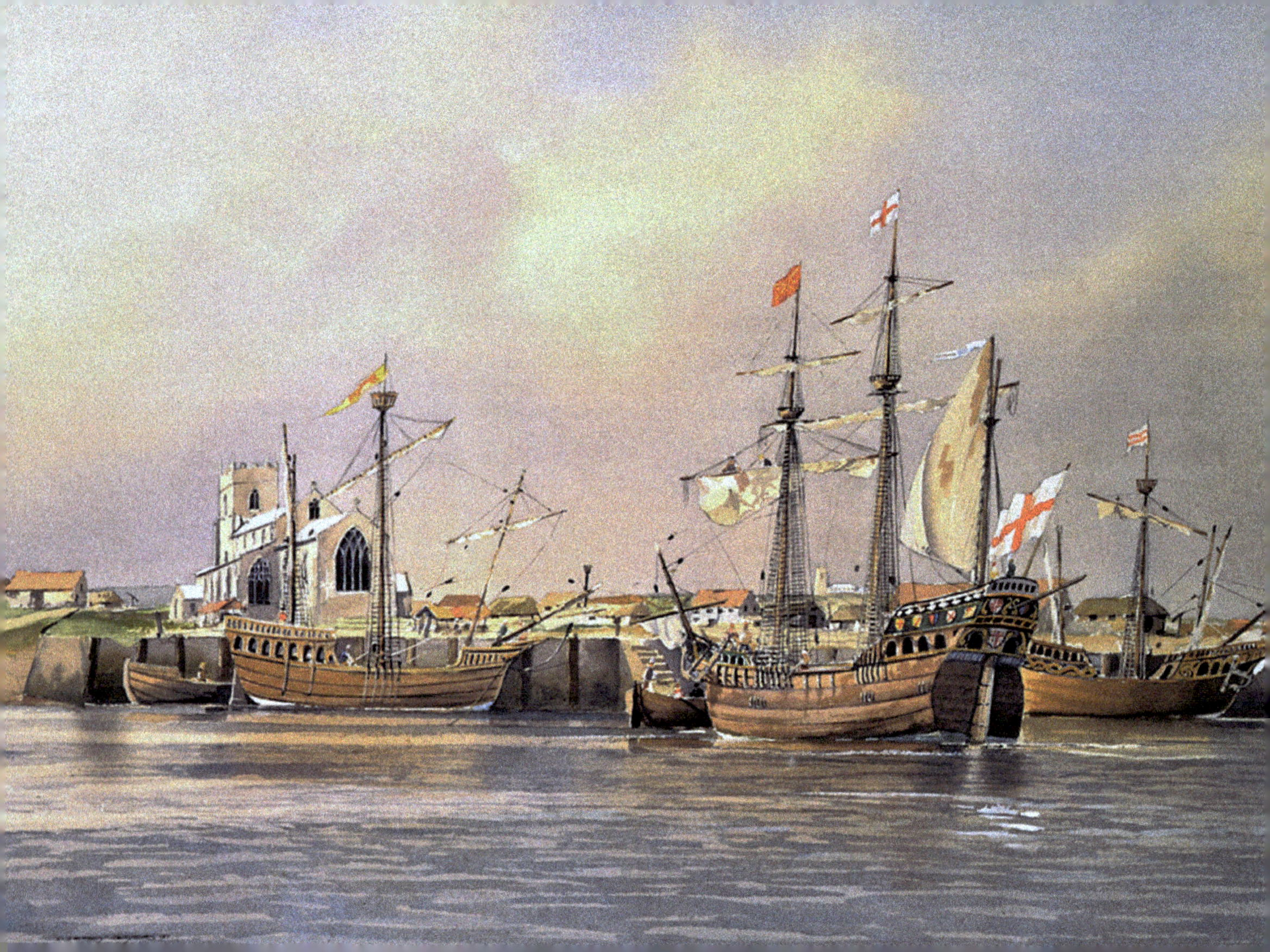

Ports and has brought Simon along for a meeting in Blakeney in a few days' time. They are staying in Wiveton so after they disembark we go with them. Soon a man pulling a small cart appears, their baggage is loaded onto it and we leave.

There are at least a dozen places in the town where they could stay, but Henryk is a regular visitor, and he prefers the Golden Goose tavern, just off the main street on Langham Road.[8] After settling in we join some of the locals in what we would call the bar, where we'll have a meal. There is only one room for this purpose and it is busy, but it is quite large. There are boards (tables) to sit at and forms (benches) to sit on. There are also three-legged stools and a cut-down barrel being used for a gaming table. By the hearth there is a high-backed settle with broadsides and such tacked to the back. There is a chest near the door that leads to the kitchen, and above it are shelves full of pewter dishes, earthenware mugs, and pitchers, all seen through a haze of smoke. Our table adjoins another and we are soon included in the conversation which seems to centre around the recent changes to the estuary. A white-bearded boatman speaks up.

---

8. Not the Langham Road of today but an older road that entered Wiveton along Chapel Lane.

'When I was young there was a lot more water than there is now, that's a-changing all the time, the sea keep a-pushin the land backards you see, thas come five a six furlongs[9] in my time and the end on it, "the Black Isle", thas moved west by another two at least and the Haven entrance with it. Thas where the town of Blakeney was for years, the whole haven got its name from that place. But that's got smaller and smaller and is nearly all gone now, just a few huts, and sheds left. Most on 'em what lived there went up to Snyterley and they now call that Blakeney, although I still call it Snyterley.'

---

9. Furlong = about an eighth of a mile.

Another old boy chips in.

'That was Sir John bloody Heydon what started the trouble we got now, he put banks up across Salthouse marsh and that caused all them streams that went up that way to block up, I was just a boy but I remember it. Then there was another one who mucked about near the main Cley channel and that's what's a-makin' the Holfleete to silt up now.'

The room suddenly goes quiet as John Braddock, the Customs Officer, appears at the doorway. He beckons Henryk over and passes him a note from a fellow merchant in Cley who wants to meet with Henryk before the meeting in Blakeney. After he leaves we sit and listen to many stories before it's time for bed.

Next morning a boy brings horses. The tide is out so we will make our way to Cley via the causeway. Before leaving we must make arrangements for a small ferry to collect us later in the afternoon as by then the tide will be in and we'll need to be brought back across the channel. Then it is off through the busy rutted street down to Wiveton Stone Bridge. The view today is very different from when we arrived. Back then it was a wide expanse of water; today it looks like two rivers divided by a bank of mud and marsh. These streams are still quite deep and fishing boats can still use them to get down to the Haven and out to sea. Crossing over the first double-arch bridge, a trackway of small beach stones leads to a second bridge[10] that spans the Holfleete. The tide has only recently retreated so the path is covered with a thin film of mud which makes it slippery, so we

---

10. The second bridge has always been thought to be built of wood; hence the name Wiveton Stone Bridge evolved to distinguish between them. But on the 1586 map it is coloured grey, the same as the Stone Bridge. Some years ago, excavating the site where it would have stood, all we uncovered was flint work. The causeway was also of stone (or stones) probably brought up by boat from the shingle ridge. It would have been quite an impressive causeway/bridge as it had to withstand considerable tidal pressure. Could that have been the Stone Bridge?

# 1586 Map of Blakeney Harbour

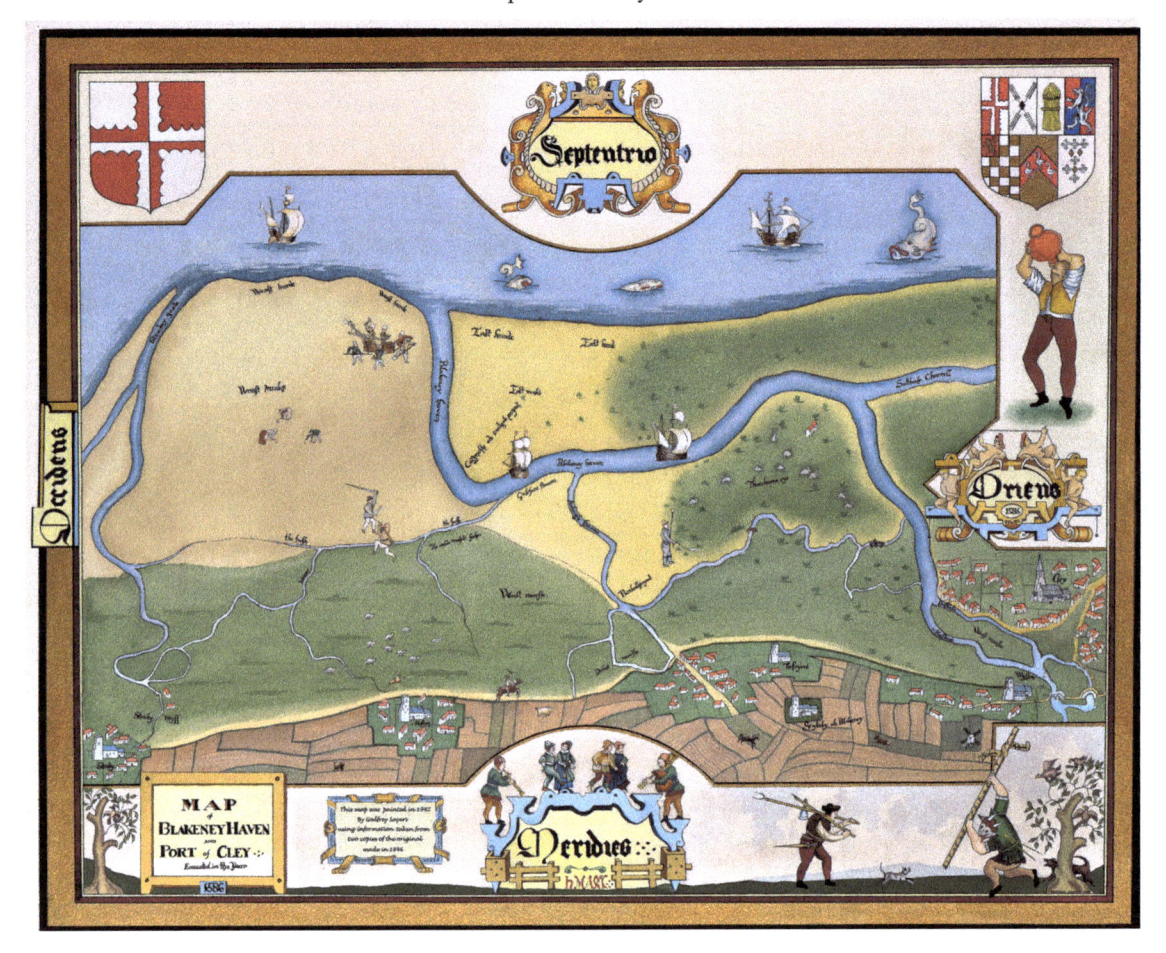

must go steadily. This is a smaller single arch bridge as it spans a narrower channel; crossing it we follow a track that leads down alongside the Holfleete to the port of Cley. All this against a background of stillness, the loudest sound the hooves of the horses; distantly the call of a boatman across the water and birds chirping in the hedges nearby. A subtle but profound difference between this world and ours. Only in the remotest parts is our world ever silent; traffic noise is the backdrop to all our lives, the silence here is deafening.

Other than the church standing above, there is nothing to tell us where we are. Today we look across Newgate Green to the Three Swallows pub, and here we have a lagoon of still water with boats of various sizes sitting or lying on the mud surrounding it. Opposite there are low quays, which have more boats tied up against them, a bit further to the north several much bigger ships are tied up. Although the tide is out men and women are working, wagons are coming and going and goods are being loaded and unloaded. Along the slope that runs down from the church fishing nets are drying on a forest of what look like goalposts.

We are to meet the merchant in the church but to get there we have a good detour around this bay to make first. Approaching we can see John Braddock has arrived before us and waits at the church door. Once more he greets Henryk before he takes us up to the small parvis room above the south porch to meet Francis Bishop where they will be undisturbed. There are a number of topics to be discussed at the meeting in Blakeney next week but top of the list is a growing concern among merchants and ship owners about the increasing number of embanking projects within the Haven. These are seriously reducing the tidal flow and increasing siltation, and are seen as the main reason that Salthouse has been so reduced as a port. What has put our friend Francis in a panic is a rumour of plans to erect a bank across the mouth of the main channels, thereby cutting both Wiveton and Cley off from the sea.[11] John Braddock tells him that this is a malicious rumour probably put about by someone who wants to deflect attention from other schemes they are planning. This does little to reassure Francis; we will no doubt discover more later in the week.

A meal is brought, and the meeting becomes something of a reunion, as Francis and Henryk have not seen each other for many years. An hour later and we are ready to leave. As we come out of the church the boats are

---

11. This was just a rumour, but sixty years later it came to pass.

lifting as the tide creeps over the mud in the bay. With the flood comes the ferryman; we can see his small boat weaving its way between the withies[12] that mark the shallows across the submerging mud flats. He comes alongside the quay to collect us and as we push off to come back the view ahead is quite magnificent. the Dolphin sits proud with the other ships and boats strung out along the quay and behind them the church. It is very difficult to imagine that one day this will just be fields with grazing sheep.

---

12. Withies: slender willow sticks.

# 5

1586

Hansa Meeting

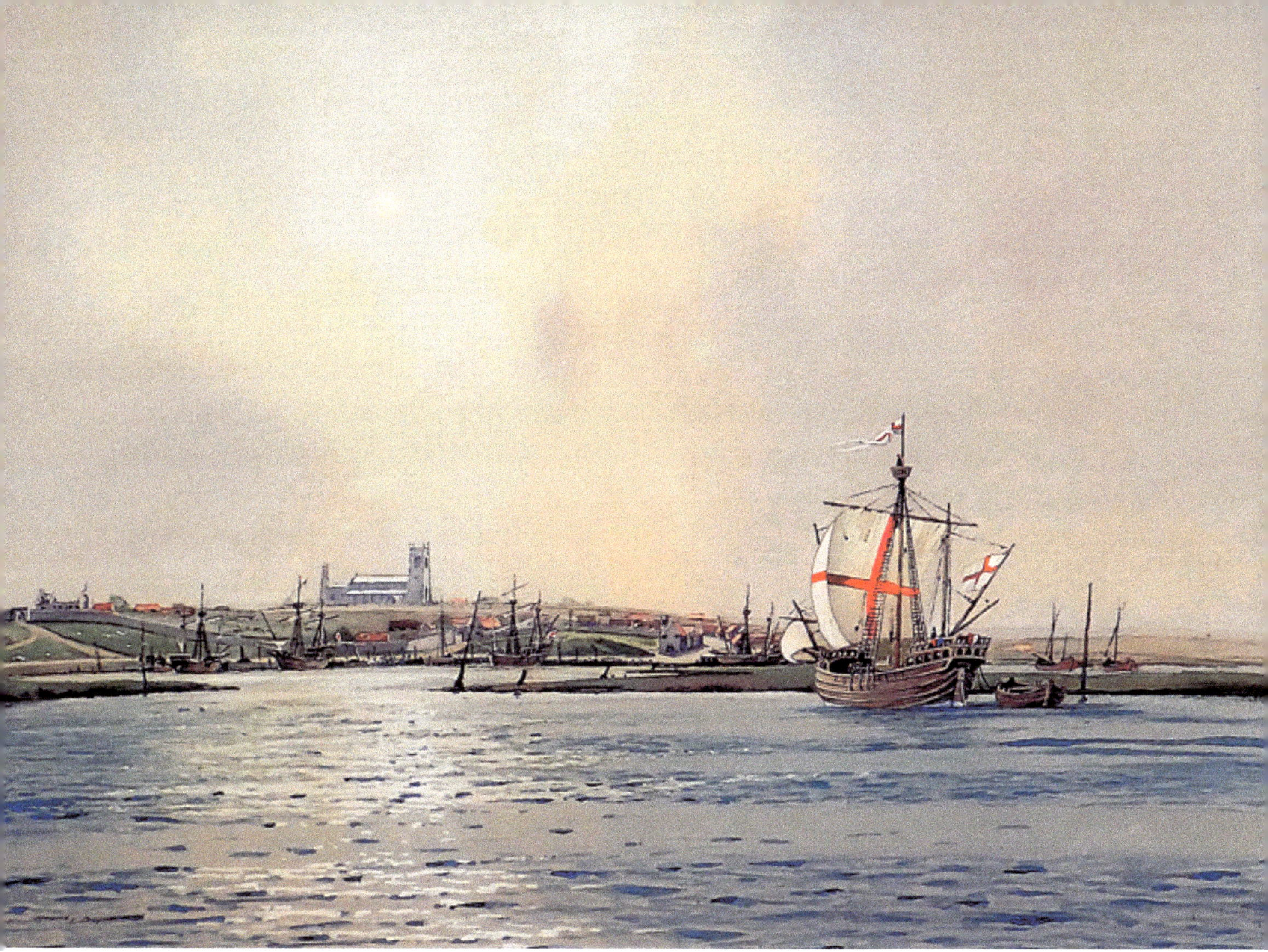

Our two merchants have had a busy time in Wiveton. The meeting down by the quay with John Braddock and the pilot has helped them understand some of the recent changes being wrought around the Haven and the concerns that the ship owners have. Useful no doubt for the when they meet in Blakeney next week. What has become clearer to them from the wealthy merchants and yeomen they have met is that just as in their own country, a new age is dawning. The Renaissance, like a distant explosion, has its echoes here, subtle refinements to both the physical and spiritual sides of life. Protestantism is in the ascendance, although Catholicism is still the religion of some older people in town and in this age of a wise Queen they are less persecuted as they might have been in the past unless, of course, they fail to attend church services; then they are fined. Simon is a Protestant but it is only recently that he has had the freedom to practise his faith in his hometown in Flanders.

He has relatives in Worstead, and he is going to visit and worship with them there. He leaves early on Saturday morning; the wind is south-westerly and a mid-morning tide means that he could have sailed round the coast and covered the last few overland miles to Worstead from a beach landing near Paston. This would have been quicker and more comfortable, but he hasn't the time to make the arrangements so he will have to go by wagon, an altogether longer and bumpier affair: the best part of a day's journey just to get there.

On his return, he and Henryk must prepare for the meeting. They have spoken to several locals and feel that the proposals they intend to put will be welcomed. The day of the meeting has arrived and they feel they are as well prepared for it as they can possibly be. A young lad has brought horses for the journey to Blakeney. Yesterday was the calm before the storm. Today, a strong south-west wind is blowing and we'll take the more sheltered shore road. We set off northwards along the main street and soon come to the small inlet we have seen many times before. A dozen small boats are drawn up onto the saltmarsh around this tiny sheltered bay. A couple of fishermen are getting their boats ready to set off on the tide, others are mending nets or working to prepare their boats for a day's work. On our left, dotted here and there, others are working the strip fields that cover the land that slopes up to Blakeney, the grand church with its two towers dominating the skyline. Our path runs above a ribbon of tangle that marks the highest tides. On our left, the curved flint wall takes us round to where we are[1] overlooking the whole spread of the Glaven estuary, which stretches inland as far as we can see. Across its wide expanse, and partly obscured by masts, we can see the now familiar cottages and storage sheds of Cley.

The estuary begins to widen as we move north.[2] On our left, Wiveton Hall: not the elegant Dutch gables of today but a simpler building built long before this time, and beyond that the partly demolished[3] Carmelite Friary. Our path follows the natural curve of ground until

---

1. Leatherpool Lane today.
2. Along Marsh Lane.
3. The recent dissolution accounting for some demolition.

SHIFTING SANDS

we are heading towards Blakeney. Distant across the saltings to the north, Thornham's Eye with its chapel buildings stands above the marsh. Closer, a wooden bridge spans a small creek known as the Gosgonge. Beyond it, the main channel has begun to fill with tide. Up on the higher ground now we see the northeast corner of the big flint wall that bounds the Friary. This will keep us company all the way to Blakeney. Ahead we can already see the masts of the ships tied up there.

Cresting a slight rise in the land, all is movement: against the racing clouds a sky full of hungry, wheeling gulls; the tide pours in, its surface darkened by the strong wind; the small fishing doggers moored along the Carnser[4] lift and swing to it. The Carnser, which separates the main channel from the mud and saltmarsh on the east side, forms a raised road leading north. A strip of dry land leads the eye to the distant Haven entrance and the dunes of the East Hurdes. Stretching east, the low shingle ridge forms the embracing arm of the Haven. The ships here are smaller than those at Cley and Wiveton because the channel

---

4. Carnser is an old name for a raised road crossing wet or boggy land. At some point, this 'road' would have been built to allow wagons to access jetties for the loading and unloading of ships along the side of the channel.

approach is narrow and winding and the quays are partly tumbled down. Yet this place is bustling: porters loading ships, fishermen pushing off into the tide and catching the sun in the distance, a ship with billowing sails makes its way through the Haven to Cley. Horses, wagons, and people abound. Ahead, a small crowd gather in front of the imposing Guildhall, which is used by Blakeney's guild of fish merchants. Its upper storey is a tavern and on its south-east corner a small turret similar to the smaller tower we saw on Blakeney Church houses a narrow stair that will take us up to the only first floor in town. We'll stay awhile and share a mug of ale.

After a short but enjoyable break we must get along to the meeting, but we'll not stride up High Street as we might today. Dodging horse dung and wagon ruts, we tread a track beaten by countless hooves and wheels, hard earth in summer, mud in winter. Cottages line its sides, smaller and fewer than today, but the village we know is there, waiting to be born. As the buildings thin out we come to the meeting house. More imposing than those it stands among, it sits at the corner of a large farm on its west side.[5] There are no more cottages on this

---

5. Beacon Cottage. This was John Wallace's home, now a BNHS cottage. He was immensely proud that he had Hansa merchant's marks in his house and was very anxious about their survival. I hope that the BHS has protected them.

side so the view to the west is open, small plots of farmed land dip toward a shallow valley to then rise again to the higher land of the esker that forms the horizon. Morston lies beyond.

We enter a low-beamed room filled with noise and smoke, almost every member sports a clay pipe. Apart from general business and reaffirming personal connections, there are two important reasons for this meeting. First is the Hanseatic League itself and how it works for the east coast ports. For decades it has been challenged by other trading coalitions. The merchants on both shores of the North Sea are keen to hold onto the benefits of their previous 'common market' by continuing to make trade deals among themselves. While relations between the League and the men of this coast have not always been the friendliest, much has changed, and this century many have thrived on the trust and friendship that has built up. Not wanting to go back to old ways there are very good reasons for wanting to reach a new trading accord. Henryk and Simon bring assurances from their fellow merchants across the German Ocean of their support for any agreement that will maintain the status quo.

The other hot topic is the local threat of embanking the saltmarshes and the problems this will bring for the Haven. Henryk begins the conversation with an apology as he can see that it is the success of his countrymen in claiming land from the sea that is the spur for it happening here. Agreement on a way forward with a trade alliance was quickly settled; the problems that will arise from embanking, while not difficult to predict, are much less easy to resolve. The fundamental problem that Henryk points out is that those who hope to embank the marshes own them; and resorting to law[6] is the only way that it might be stopped.

---

6. In 1637 a bank was put across the Glaven by Lord Calthorpe, the landowner, and a protracted lawsuit duly followed. It was eventually successful but took so long that the channels to Cley and Wiveton became so silted that their role as ports was over anyway.

Business over, it's time to go. Before we leave, Simon, who is here for the first time, takes out his knife and cuts his merchant's mark into the large beam over the fireplace. A local merchant, Tomas Starling, whom Henryk has known for many years, invites us all to his home for a meal. He lives in Morston, so a slight detour to the west before we set off for Wiveton is in order. The boy is waiting with the horses and we leave immediately. We head off along a track[7] through the farm; this joins another coming from the west end of the village and heads off uphill toward Morston. Reaching the top, we have a magnificent view. To the north and east a

---

7. Little Lane today.

great sheet of water, its waves white-topped by the strong wind and many small craft making their way to and from the towns of the Haven. A final half mile and the road drops down to a tide filled inlet overlooked by Morston's church, small boats are moored along its west bank. We cross via a short causeway leading to a bridge and on into the village. The coastal track that leads through it has a few scattered cottages either side, Morston isn't a port like the other towns of the Haven, but the small creek that leads to it is deep enough for small fishing boats and other flat-bottomed craft to navigate. These ferry grain and other locally grown produce to the ships in the outer Haven.

Tomas's wife welcomes us and a table is prepared for a now quite late midday meal. A rather grand spread awaits. In this world, you are defined by what you eat, the difference between classes important to the order of society, so great efforts were made to mark the divisions between them with laws that limited spending on food to ensure that nobody ate above their station. For those we saw working the fields earlier such laws were irrelevant; labourers could not afford much more than pottage – the staple dish – and you could eat as much of that as your meagre means would allow. The rich ate pottage too, but instead of what was just cabbage soup with some barley or oats – and a sniff of bacon if you were lucky – a nobleman's pottage might contain almonds, ginger and saffron, as well as wine. We won't be getting that, but Henryk and Simon are wealthy merchants, as is our host, so this welcome meal should be appetising.

Sadly, there is little time for socialising so as soon as the meal is finished we must leave for Wiveton. With the wind still hard from the south-west we'll have a little help pushing the horses up Blakeney Hill. At the top a very different view, from the spot where three centuries hence one of the Long family will pounce on a Customs man. A wide basin cups at its centre a huddle of cottages, a cluster of masts and overseeing all, the big church. We will retrace our steps through Beacon Cottage farm to join the town street where we left it.

―――――

With the afternoon running out we'll take a quicker way back to Wiveton. Beyond the street's cottages we are once again in an open landscape, revealing on our left the corner of the Friary's south wall, and between it and the church the Swanmarket, the one part of the village that looks more to those working the strip fields than to the fishermen. And the great church, a statement of power and wealth that today is roaring its dominance in the high wind.

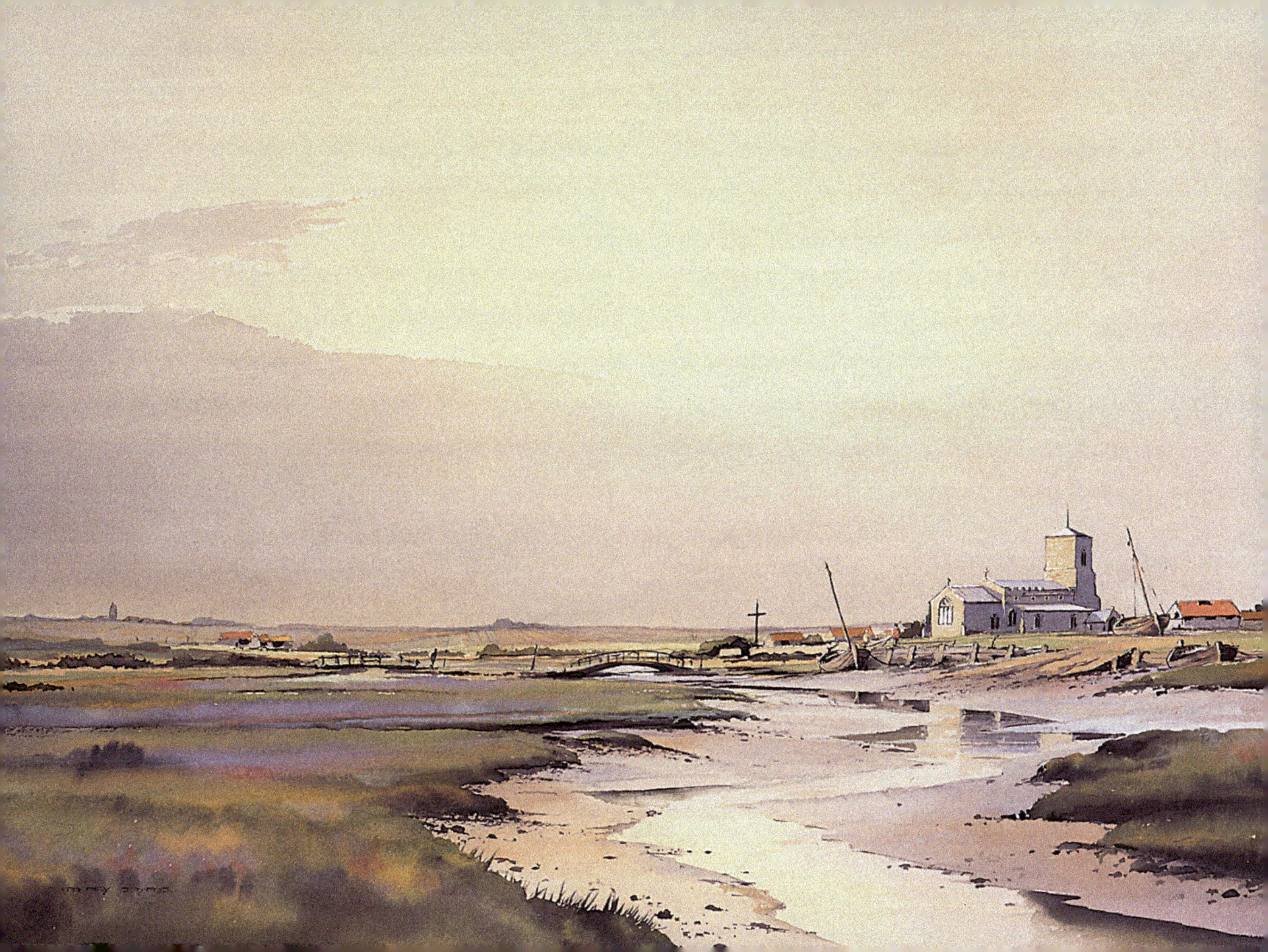

Further on the land drops away, hedgeless strip fields on either side. We are buffeted by a wind that blows hard in this open land, that rolls off to the treeless horizon. Centre stage, the towns of Cley and Wiveton face each other across a wide tide-filled estuary. A little down the hill we pull over to allow a loaded wagon to pass. The manes of the horses stream in the wind, the wagoner holding the reins with one hand and his hat with the other, smiles his thanks. He is taking flour to Blakeney from the windmill atop the rise ahead of us it marks the entrance to Wiveton. Just past it we come to a crossroads: turn right to go to Langham, straight ahead for the marketplace, Wiveton

church and its quayside, and left for the Golden Goose. We turn left.

When we reach the Goose, the ferryman, who it seems has been waiting a long time, hands Henryk a note telling him that the ship's master dropped the Dolphin down to deeper water as the tide was smaller than expected, being held out by the strong offshore wind. He wants to sail on the first of tomorrow's tide, so if we are ready to depart the ferryman will take us down to the Dolphin. After a long day this is not ideal, but neither Henryk nor Simon has anything to keep them here, so we get ready to leave. With everything packed the porter who brought their luggage to the Goose is here to take it back. He says that the tide has almost gone but we'll have the last of it to take us down river and by the time Henryk, Simon and their luggage are aboard ship he'll have the first of the evening flood to carry him and us back.

Arriving at the quayside, it is mostly mud that we can see. The ebb is almost at an end, but the channel is deep so there'll be water enough. With so much wind there is no need for the sail, the mast and the ferryman standing up in the stern catch enough wind to send us along at a good pace. The Dolphin lies in deeper water where the two channels meet.

Coming alongside, we grab a rope ladder to clamber aboard. The crew pull the baggage up on ropes and secure the cargo ready for departure tomorrow. But as we don't want to end up in Ghent, we'll say goodbye to our Dutch friends, climb aboard our time machine and come back to 2020.

# 6

1845

Coastal Harbours Commissioner's Visit

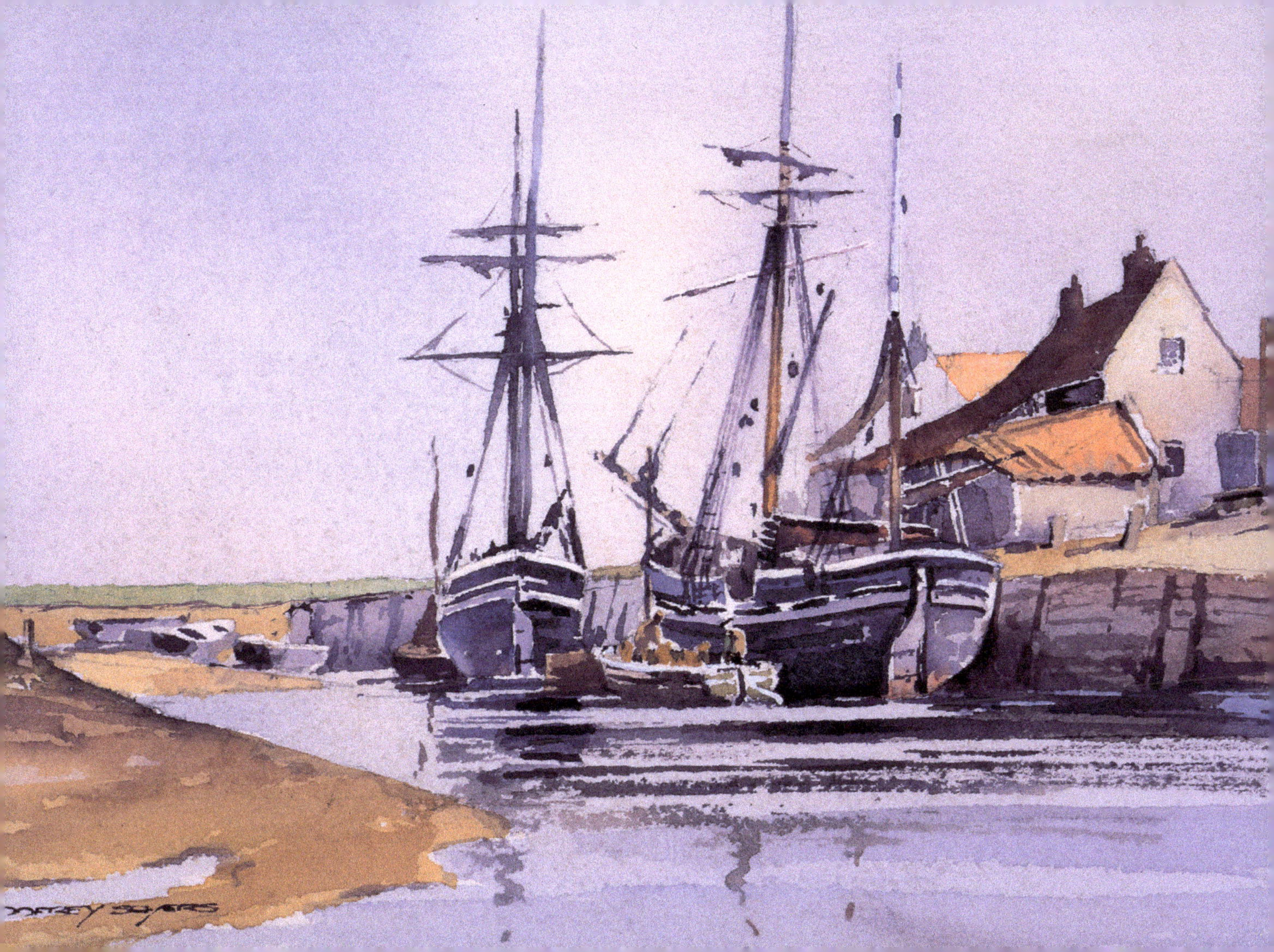

BLAKENEY  CLEY

WIVETON  SALTHOUSE

Three hundred years of wind, tide and men with spades have transformed Blakeney Haven. The spit has moved west by almost half a mile and the harbour entrance curls inward forming a ridge on which the newly built Lifeboat Station stands.[1] In 1824, a bank was erected across the Glaven cutting the port of Wiveton and most of the quays at Cley off from the sea. What is left of the channel to Cley is badly silted and only reaches the much smaller length of quay that remains north of the embankment leaving Blakeney as the main port. Just a few years after our last visit the Blakeney-Cley bank was erected, enclosing what is now Blakeney Freshes. Further embanking and the landward

---

1. This the original Lifeboat Station was built in 1824 and stands behind its replacement that was built in 1898.

march of the shingle ridge has severed Salthouse's link to Haven. So many areas of saltmarsh have been claimed and the tidal prism so reduced that Blakeney Haven is now just a shadow of the great refuge it once was. The selfishness of the landowners responsible seems very callous today but then so do the Highland Clearances; it was a different world with different rules back then. The reduced outflow of the tide from the harbour has accelerated the growth of the Point westwards and the burden of sand brought in from sea, along with silt coming down the Stiffkey and Glaven rivers, has reduced the depth and width of the channels.

The embanking at Cley was not a straightforward fait accompli, although in the end, that is what it was. Construction of the bank had already begun when concerns were raised about the damage it would do to the harbour by reducing the tidal outflow. An engineer of no less status than Thomas Telford was brought in. His brief was 'to survey the marshes to be embanked and drained, and of the works already carried out, and report as to the propriety of going on with the present works, or of adopting some other plan for effecting the proposed embankment and drainage, and to deliver an estimate of the expense. 'In determining upon the plan to be adopted, 'Mr. Telford will please to take into consideration the effect which it will have upon the harbour of Cley, any injury to which must be avoided, and the improvement thereof is an object, <u>so far as it can be effected without any material addition to the expense</u>.'

In his report, Thomas Telford made very clear the damage the proposed bank across the Glaven would do. Based on this, his recommendation was that the river should be embanked on either side as far as Glandford so as to maintain some tidal flow to flush the harbour. His report was rejected as it did not conform to the underlined words above. Had he read them he might well have saved himself a lot of trouble, although of course he would not then have been paid. The bank was built.

It is into this scenario that we will venture and accompany an Examining Commissioner from the Commission into Tidal Harbours. The date is 1845. Before that, in 1817 an Act was passed here that allowed the creation of the Blakeney Harbour Company, a group of well-intentioned but also vested interest parties, which undertook major improvements to the harbour.

We join a group of local townsmen who are taking the Examining Commissioner on a tour of the inner harbour to show him the 'improvements' they believe they have made since the Company was formed. We begin by walking along the sea bank towards Morston, half-way along and near to the old sheep bridge; the inspector is invited to look at a dam that has been placed across the creek.[2]

'Why have you done this?' he asks.

The Harbour Master replies. 'Well, you see there's already a low bank circling the eastern end a these marshes, that was put up in past times to try an make all the water that covered 'em run out the same way as it come in, but we think now that a lotta the water was gettin' away through this crick to Morston, an takin away what was needed to keep it clear.'

'Well,' says the inspector, 'I think you have wasted time and money, you see by stopping the flow through this creek you have created dead water at high tide, so more sediment will drop and the creek will silt up more quickly.' Faces fall, not a good start.

Making our way back to see more of this grand project we come off the sea bank into a

---

2. Agar Creek.

lumber yard; large stands of timber all around, a storage barn on our right,[3] another on our left. These belong to Gus Hill who lives in Red House which lies just around the corner. He also owns the quay that fronts this yard and is of course with us on our tour today.

'There are masts sticking up above that barn', one of us says, pointing north.

'Yes, that's the Mary Anne', Gus tells us. 'There's a dry dock there along the north a my land. They put ships in there on Spring tides and so they can work on 'em. They caulk 'em and tar 'em and then float 'em out when the next Springs come around, or a bit sooner sometimes if the work is done and there's a big enough tide.'

We go on to find our way is almost blocked, the space ahead littered with the stuff from the sea: nets, ropes, buoys and boats, anchors, chains and fishing gear cover the area. We must weave a way through it all to get to the road. Even then the assault course isn't finished, the roadway is filled with horses and wagons. It is clear that Blakeney has moved into a golden age.

---

3. North Granary.

Remarking on how busy it is we are told that the grain-out, coal-in trade is flourishing, as is the fishery, especially oyster fishing; 300 vessels are registered from Blakeney. We are here on what is probably the busiest day of summer and from what we already know, perhaps all time. Ships are tied up all the way along, and more than thirty, four-horse wagons bringing grain are lined up unloading. Others are arriving in Westgate Street; some have come from as far away as Dereham. Apart from coal, stores and groceries also come in by sea. Some supplies are brought twice weekly from Norwich by the carrier, John Miller, but he could never bring in enough for an entire village. This is a busy town.

A lighter is being swung on the tide to take its place alongside an earlier arrival, a crane at the end of the quay is swinging bags of grain into the hold of one ship, while further along others are discharging coal into the emptied wagons. At our backs we have a granary where some of the surplus grain will be stored, a bit further along a row of cottages, and tucked behind them a maltings and a pub called the Barking Dicky. These cottages have raised ledges along their fronts[4] on which some children are playing; these are quite wide, and one is lying

---

4. When these cottages were built it was obviously understood that they would have to withstand inundation by high tides.

flat on one. Young boys scurry about with little handcarts collecting small loads of coal to take to housewives in the village, and men, backs bent, trudge back and forth across the road. It is a noisy, dusty place. We pause to watch the coal being unloaded into the winter store. A more laborious task is difficult to imagine; I'm sure many slaves had an easier time. Peering over the quay we can see two men in a ship's hold filling heavy bags. These they lift onto the porter's back with a device like a stretcher. The coal is getting low in the hold and the tide is out so the porter has to climb a ladder up onto the deck, then another onto the quay, he then has to cross the road and climb up more steps to tip his bag into the coal house. All for less than a penny a ton. The power of this place is overwhelming . Seasoned as we are to the constant roar of traffic, it seems almost silent, just a soft hum of physical exertion: creaking pulleys, snorting horse, laughing children and dust. To our noses, salty mud, horse dung, sweat and tar, overshadowed by looming masts, this world is painted a very different colour to the one we know.

An elderly man with a white beard approaches pushing his own cart. His is a slightly different mission. He tells us that he goes around Blakeney and other nearby villages selling local produce, most of it grown himself. A brace of rabbits and some wild duck hang on the side, and in the cart butts,[5] caught in the creek, plus cockles, winkles and samphire. We ask him if the coal delivery boys get paid, and he says they get ships biscuits in return. If this is true it might account for the terrible state of some of their teeth. Perhaps the women they deliver to give them something more to eat. We show our surprise at the number of ships along the quay; he tells us that there are at least another dozen more anchored in the Pit waiting to come up. Some of these are on tight schedules and will have their coal and other cargoes brought up by lighters, like the one we saw earlier.

---

5. Butts: flounders.

Reaching the end of the quay, the coal barns stretch up the side of High Street and across the road the Guild Hall where we broke our journey on our last visit. Its use has changed, it now belongs to the poor and is capable of holding sixty cauldrons of coal, and from this use brings in £4.4s per annum for the poor of the village.

Once more we make our way along the narrow Carnser. Saltmarsh and a muddy creek one side, the channel on the other. Dozens of mussel canoes are strung out along the waters edge and fishermen are painting and tarring them ready for the mussels' season that's soon to come. We hear that the mussels are cultivated in the harbour, and the small seed from which they grow is brought in by fishing smacks from Boston. One of the fishermen, John Baines, tells us of a member of the crew of one of these smacks called The Skylark, who is a very good singer and is in a Blakeney choir. If they have an event when he isn't in port, he has to travel overland to attend. Throughout the centuries of Blakeney Haven's maritime trade there has been this interchange and movement of people: a Blakeney man might marry a Newcastle girl, or a Blakeney woman might marry a visiting seaman; both men and women migrated.[6]

6. My grandmother was born on North Shields.

How different this coastal strip is to the hinterland. For centuries the inhabitants of inland villages would, in all meanings of the word, only have had regular intercourse with their own and nearby villages and only married and had children with those nearby. Here there would always have been people coming and going and staying and leaving, its legacy is a rich genetic mix. And in times of famine nobody along this coastal strip would have starved; even without Piddy Palmer and his little cart, they would have foraged for his wares for themselves.

On the horizon something that is very familiar to us time travellers, but perhaps more eye-catching to some of those we are with: the Watch House,[7] just ten years old. In the middle distance on the other side of the water, and still looking very raw, are the huge piles of silt and mud excavated from the New Cut.[8] This deep, new stretch of water has brought an immediate (if not long-term) improvement. Gus tells the inspector how it was done.

'You see before we did this the channel went out in a big bend to the west and that was difficult for ships to get around, this here straightening has made it much better. Now they can come straight up', he points, 'we put them big old posts in every little ways so when there's no wind they can pull themselves in with their windlasses.

Once again the inspector is not impressed, saying 'rivers and tidal channels have a natural tendency to meander.'

---

7. The Watch House. Originally it was called the Preventive House as the forerunners of Customs & Excise used it to prevent smuggling. But it was spectacularly unsuccessful in this role, as the smugglers used it as a store for their contraband. So after just eleven years, it was taken over by the Coast Guard and became known as the Detachment (as in detached from the Coastguard Station in Morston), and it was used for foul-weather watch; two men would be posted there when the weather was bad on lookout for ships in distress. It had a magnificent flagpole up which a flag would be run up if such distress occurred in daylight. A signalling lamp would be used at night.
8. The New Cut was dug in 1817. Before that time, the channel wound its way through the marsh with a large bend extending to the west of today's New Cut.

'You can see already', he says 'how the tide's natural inclination towards a winding channel is depositing silt along its edges: adding that they 'would have had more success if they had dredged the existing curved channel'.

More glum faces. But they are struggling to make the best of what fate and greedy landowners have handed them. We learn from our disheartened companions that the inhabitants have struggled for nearly half a century to prevent enclosure of the tidal lands by adjoining landowners, even at the expense of many suits at law; but all these efforts were frustrated, and now 900 acres have been enclosed to the great detriment of the harbour. John Lee, a merchant, said that in 1820 ships drawing nine feet were able to get to the quays at Cley, but since the bank had been in place that had been reduced to just five. So a further contemptuous development in the form of a second bank across the Glaven, just seven years after all this work had been done, rubs salt into an already open wound. But none of this influences the inspector's view.

'This was a harbour open and free for all', he says. 'Now not only has a public harbour been destroyed by the encroachment of its tidal lands, but it has also become the property of a

private company who largely profit therefrom.'
Not a good news day.

Returning towards the splay of the High Street we see a wagon being guided into position by the granary on the right. The hill is so steep that the horses pull the wagons up under the gantry and large blocks are put behind the wheels to take the weight. Opposite are the offices of Page & Turner, the big ship owners and traders. They have ten ships, and offices in Wells and Burnham Market as well as Blakeney. We are invited to go up the steep steps to the office. A ship has recently come ashore by the Long Hills in rough weather. Page & Turner after (we suppose) not offering to refloat it have bought it. The ship is the Sir John Colomb. She isn't badly damaged and Mr Turner has arrived to arrange for a boat to take a carpenter out to begin repairs. Once this is done she will be refloated on the next big tide, brought up to the quay and converted for Page & Turner's use. They have acquired at least two ships this way.

Mr Page offers to take us to lunch in the White Horse Hotel just across the road. The bar is not full; those we've seen toiling along the quayside will come here later. He tells us a little of the history of his company. From the earliest days the basic trade was coal-in, grain-out, supplemented by the basic necessities of life. His company is flourishing, sailing packets to London and Hull once a fortnight and trade, as we saw, is doing well, so for the moment the future looks rosy. But Page is an astute man. He knows that Brunel has completed a railway from London to Bridgewater and others are being planned. Just last year, a line was built from Norwich to Yarmouth, so he can see that moving large amounts of coal and grain by rail could eventually mean a reduction in coastal trade with small ships.[9]

It's time to move on as we are meeting Ted Walker by the Lobby[10] who is to take us to Cley in his two-horse wagon. Ted is the carrier who goes to Norwich once a week, a day to get there, and another to get back, he tells us. Although the New Road is now well established we take the old shore road as it's easier for the horses. Just a short way along, opposite Manor Farm[11] we are surprised to see a well made-up road leading out to the north. This, Ted tells us, goes across the fresh marsh to the sea bank, over it, and continues on the saltmarsh where it reaches a shallow (at

---

**9.** Eventually, toward the end of the century, Page & Turner would move their business to Holt to take advantage of the railway.

**10.** Now the ice-cream shop.
**11.** Now the Manor Hotel.

low tide) stretch of firm sand that fords Cley Channel. It continues over the marsh on the other side until it reaches the shingle ridge where it turns west and follows it all the way to the Point and the Pit side, where ships unload into carts. This is a very busy place that brooks little delay.

The old road takes us past Friary Hills and Wiveton Hall and round to the junction with New Road at the bottom of the steep hill. We turn left to follow the New Road as it runs along the south side of the new embankment.[12]

Over the flat grass fields are the broken remnants of Cley's old quays that after 800 years will not see a another tide,[13] sinking into the past to join the cobs and doggers that once tied up to them. We have arrived in Cley. Along the High Street we can see the odd mast above the houses on our left, but arriving at the Quay it seems oddly strange. The mill is facing in an unfamiliar direction, its appearance drab without its fancy white sails. In the afternoon

---

12. Before this road was built, the only crossing at this point was a small bridge covered at high tide. In 1816, a group of five people returning from Cley by boat were drowned when a rope fouled. No doubt this added weight to the argument for the bank and accompanying road.
13. If they had remained they would have seen just one more tide in 1953, when a tidal surge did what litigation couldn't, and flattened the hated bank.

sunshine, the quayside is overshadowed by a large three-storey granary standing foursquare to the quay: a monument to Cley's former dominance as the Haven's premier port. Now the extension of Blakeney Point, and the rapidly increased siltation has reduced Cley channel to a state where only small ships and lighters are able to get up to and turn at the quay. Nevertheless, the quayside is busy, wagons are coming and going, a couple of lighters are being unloaded, and a crowd has gathered for a revivalist meeting. Between hymns, members of the congregation stand and make pledges of faith. A man whipped up by the occasion leaps up enthusiastically, and after listing all the inconvenient situations he can think of in which his faith would be tested, concludes with the claim that should 'they' (not sure who) seal him in a barrel he would shout glory through the bung hole? Time perhaps to come home.

# 7

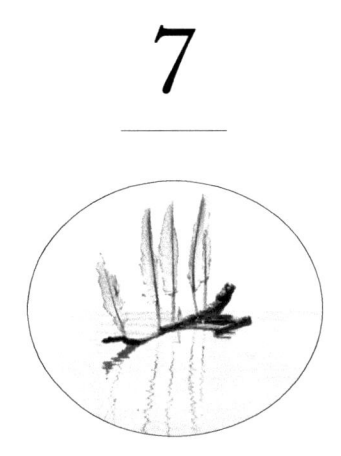

## 1945

## A Blakeney Child's Memories

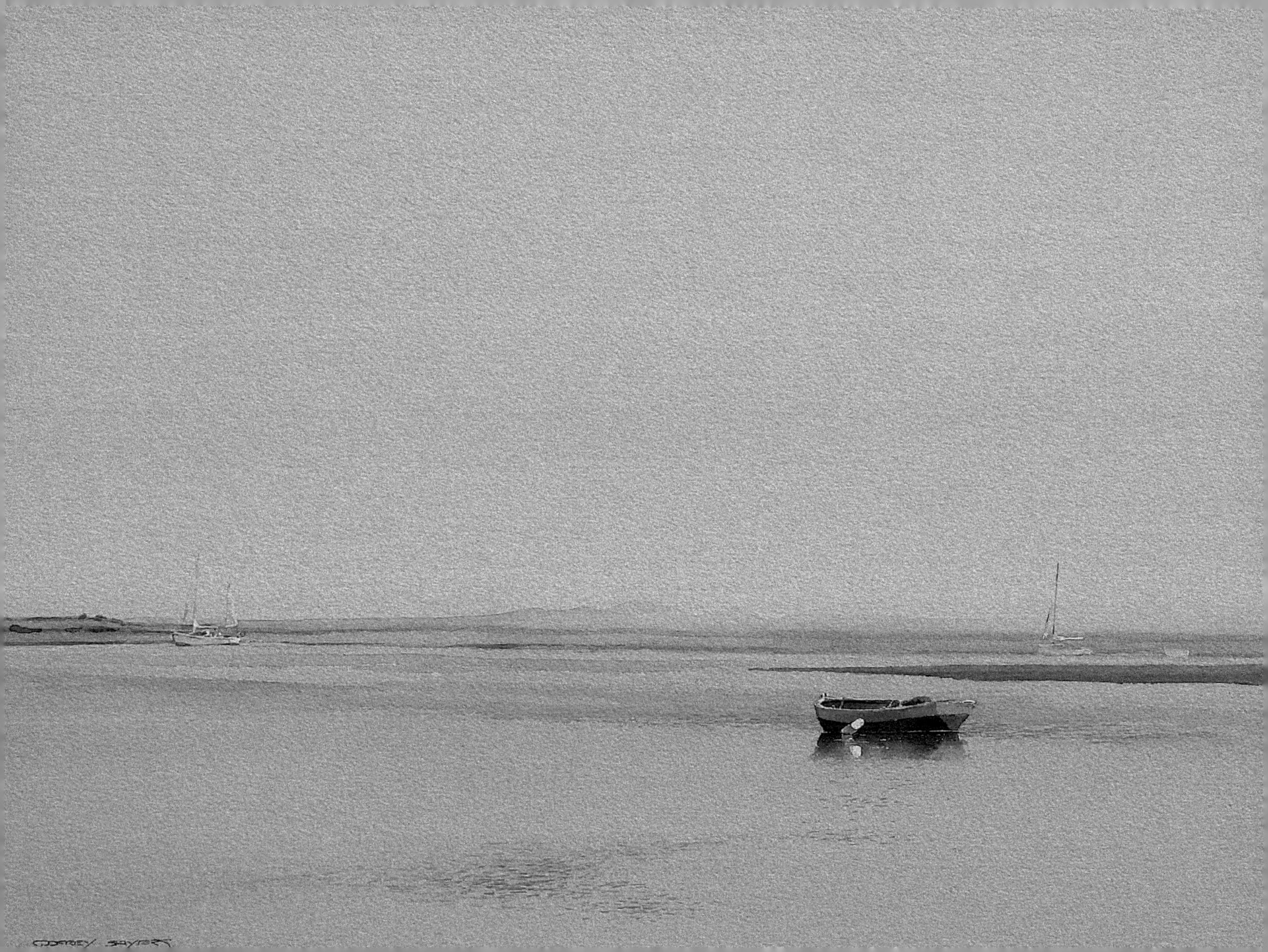

Blakeney  Cley
Wiveton  Salthouse

It is 1947 and Blakeney is enjoying a few years of tranquility, poised between its hectic days as a port and a new life as a destination for holidays, recreation and retirement. Geophysics has continued to count down Blakeney's demise. Not as a port, (that is well and truly over) but also as a harbour. The shingle ridge had rolled landward backfilling the channel to Cley, requiring another new cut to be dug. Siltation caused by centuries of greed and mismanagement sealed not just the fate of Cley and any viable channel to it but is slowly doing the same for Blakeney. The growth of Blakeney Point has continued apace, its reach only a little less than today. But it is soon to meet with the first ever check on its inexorable westward progress, the 1953 tidal surge. So powerful that in just a couple of hours most of the Point's newly developed sand dunes and sandbars were washed away. Siltation, already rapid, was then given a further boost by the interference of one Professor Francis Oliver. A well-meaning man, no doubt, but seeming to lack any understanding or concern for the consequences of his actions.[1] He had experimented with planting spartina and other cordgrasses in a number of harbours around the United Kingdom, with little success. But Blakeney mud suited spartina to a T, and it hasn't looked back. Smothering the mud flats and raising

---

1. That is giving him the benefit of the doubt.

them above the mean high-water mark, thereby compressing a century of siltation into a few decades. It's a sad irony that just as Blakeney moves into a new era of activity, so it is also reaching the end of any long-term viability as a harbour.

On this visit we will see Blakeney through the eyes of a small boy brought here from the city in the aftermath of the Second World War.

'Everything I saw in Blakeney was unlike the urban neatness of Kingston-upon Thames. Although, it was not just what I saw, equally powerful was what I could smell. The washing strung across my grandmother's yard, the wallflowers that grew on the tops of the walls surrounding it, her outhouse leaking traces of ash and soap, and her cottage. Inside it was a rich, almost indecipherable mix of scents, a blend of substances that had had ages to soak into the fabric of that little place, few of them in general use today. Coal, candles, paraffin, fuels for softer warmer lights, light only then going out in favour of brighter, instant electricity. During the days of my childhood, I discovered that all the cottages I went into smelled the same. It was a composite smell, like flower shops and tobacconists. The most unfamiliar mix of scents was to come a little later when I wandered down to the bottom of Westgate Street and looked on the quay for the first time. The breath of the sea, of mud, marsh and the last spring's tide line, and, lingering still as part of an older alchemy, tar. The ships and smacks that had been sealed with it were gone, but enough tar-encrusted detritus remained rotting into the mud of Low Quay to offer my keen four-year-old's senses the chance, just once, to touch directly a world that had gone. Like seeing that elusive green flash as the limb of the sun drops below the horizon. I am in no doubt that it was that brief touch that gave me my fascination with Blakeney's history.'

'On the very first day wandering up Westgate Street I met a boy of my age who had also just moved to the village. He lived in the last row of cottages called West Corner,[2] at the top of Westgate Street. He approached and timidly introduced himself, we became firm friends, and for the next few years were inseparable. And so my new life in Blakeney began. Together we had adventures and a freedom quite unknown to youngsters today. We began to explore the village quickly becoming familiar with every passageway, loke and yard. In those days nobody seemed to question why you were anywhere, or what you were doing there. Almost every big yard had a well in it. These were a fascination and we gauged the depth of each by dropping stones. It amazed me how long it could take for them to hit the water, and the further we went up the High Street the deeper they became. Some of those at the highest part of the village were more than a hundred feet deep, and it seemed you could have run a hundred yards in the time it took for a stone to splash. The very best places to explore, when we could get into them, were the old disused granaries and coal stores. Although very run down and neglected, they were as they had been when they were

---

2. Now Ladybird Cottage.

in use. The old wooden steps and doors were still there, as were most of the interior fittings such as pulley wheels and wooden rollers, some almost worn through where ropes had run over them. In one granary there was an intact donkey-driven bone mill, two enormous grinding stones with a long wooden shaft onto which the donkey was harnessed. We didn't know that. We thought it was some sort of punishment device that convicted prisoners had to drag around.'

'What we could not have guessed was that as well as animal bones those of dead prisoners and soldiers were also grist for its mill. The bones had come in on ships, the traffic reaching a peak during the aftermath of the Napoleonic wars and were ground up to be spread on the land as fertilizer. For a time this part of the country was dubbed Shangri-La because the indigenous population lived so long. Apparently, one prerequisite for longevity is to have plenty of selenium in your diet. Bones are rich in it, so who knows? At that time only two of the barns had been converted for other uses. One had been made into the Blakeney Hotel ballroom and the other, the long barn formerly known as Ditchell's, which adjoins a boat shed and stands with its gable end facing Low Quay, had become a dwelling.'

'It was playing in and around one of these barns, the one known today as North Granary, that we came upon the place that was to become our launch pad for grander exploration. It was once Gus Hill's timber yard and alongside a muddy basin that had been a dry dock. My grandmother always referred to it as the Mash Green. I thought this was her Norfolk pronunciation of "Marsh Green", but I now understand that 'mash' may be an old English word for land with a particular proximity to the sea, particularly where it provides access to a marsh path or grazing track. Along the coast, many of these have stands of fennel (a plant used in cooking shellfish) just where you come off the marsh, planted long ago for those coming home with their supper. In those days the Mash Green was a rubbish tip, a dump for the village but a treasure house for us. Among the rubbish, we found the makings of a world of our own. The area was surrounded by rough grass and huge stands of Horse Pepper (Alexander's), the hollow stems of which are in tubular sections like bamboo. This made them usable for all sorts of things – high-powered peashooters, telescopes that made the world appear small and fuzzy, pipes for smoking old cigarette ends which made the view through the telescopes even fuzzier, and the hulls of ships.'

'If we pulled the stalks up completely we could shape the root into a prow and the rest would remain more or less watertight. Feathers became sails – white goose feathers were especially sought after – but if they were scarce there was no shortage of others. A row of four or five of these stuck in along the top, a bit of broken glass pushed into the bottom for a keel, and we had as fine and functional a model ship as any boy could wish for. They only took a few minutes to make, so if one did not perform well a replacement was quickly made. By twisting the feathers to different angles they could be adjusted to sail into as well as off the wind. More ambitious models had outriggers made of smaller stalks to make them sail faster.

Sometimes we would make a dozen or more and set them off together at high tide, and despite having been fashioned from rubbish they were an elegant sight as they bobbed and weaved their way along the front of the quayside. Years later when I acquired a sailing dinghy and found I could sail it instinctively, I realised I had done my sail training back in those days of Horse Pepper clipper ships.'

'It occurs to me now that the idea for those little ships had come from the older boys. I don't think we invented them for ourselves, so it seems likely that the design was being passed down from generation to generation but only through the children. Other little ships may once have bobbed along the quayside, leaning over in the same breezes that brought coal and other exotic cargoes to Blakeney in its medieval heyday. Sadly, we were the last receivers of those oral packages. No other children came after us to play on the marshes. The same broom that had begun tidying up the old granaries also swept away those last small traces of Blakeney's days of sail.'

'The last ship to come up to Blakeney did so just after the First World War, and the village had been frozen in time since then. Going into the old granaries looking for matchboxes, we found some that would have held matches that lit the cigarettes of men that died in that War: they were as they had been when that last ship came up to the quay. There was no mains sewage and electricity only recently arrived. The first film I ever saw was 'Treasure Island' at the cinema in Holt: The Picture Bus. A bus from Pye's Garage took us once a week. Coming back, I got off it at the top of Westgate Street, a dark winter night, no street lights or lights of any kind that I can recall, except those that illuminated the King's Arms pub sign, hanging out over the road looking exactly like that of The Admiral Benbow. I had stepped out of a film and then back into it again. Just two cars in the village, the doctor's and Brigadier Ropes'.

On a Sunday we'd sit on the old granary steps at the bottom of Westgate Street collecting car numbers. Not Saturday, Sunday was the only day there were enough cars to count. Even then we would only get as many as there were lines on one page of an exercise book.'

The ways of the village had changed little either, if we reached back into the last chapter, grabbed John Baines and took him down to the New Cut and he would see nothing different, the same oilskins, sou'westers and woollen mittens: tough, dark-faced men. 'Will Watch', seventy-years old, grim, weather-beaten, tough, one arm and blue powder burns splashed across his face,[3] was every inch a Henry Morgan. John would of course have seen familial traces, resemblances to those he had worked with, and if he in turn could go back he would have seen combinations of them weaving back down through the centuries, perhaps to be among the faces of those Gyles and James Dobbe chatted with so long ago. A lineage that ended in the twentieth century, as the Point's inexorable migration westwards covered all the mussel grounds with sand. Apart from 'Tippney' and 'Bugle' Baines, all the mussel fishermen in Blakeney were Longs, and

---

3. His arm was blown off in a punt gun accident.

apart from young Billy were all well over sixty – Will Watch, Billy's father, George, Sammy, Freddie and Fatty, men who had rowed the Blakeney lifeboat, fought in two World Wars and inhabited a place that hung back in the past, unsure of what the modern world held, as if putting a lie to the courage these men had shown all their lives. A tough, hard lot that gave no quarter and expected none. To expect that from the sea, that it would give leeway, would be like believing in fairies.

The mussels were brought up from 'below' (the lower harbour) in flat-bottomed boats called canoes, a longer row than in earlier times due to the Point's westward growth. No outboard motors to help you, the canoes were either rowed or punted with an oar. The oars were enormous, but Will Watch could pull two together, their ends gripped in his one huge hand. As in the days of John Baines, mussels were brought round as seed from Boston every three years or so and sown in the lower harbour. They took a further three years or so to mature depending on where they were sown. Summers were spent thinning them out as they grew. A hand-net and rake were used to gather them on the dry; a large, hooped rake with a net attached and a twelve-foot handle called a whim, in deeper water. Tools unchanged for centuries. Before leaving to go down on a winter's day they would soak their woollen mittens in the channel as it made them warmer, although warmth in that climate is relative.

Change when it came had to force its way in, the first mechanical mussel-grader appeared in other places years before the men at Blakeney could be persuaded to even look at one. Even then they had to go through a brief transition period of using primitive, inefficient hand-cranked versions before getting to those with engines. At the time it seemed like Luddism, but it was more likely to be inertia, as if the hands of all their forebears were hanging onto their sou'westers holding them back in the world they knew. To be fair, the change was sudden; in the space of ten years life here moved from the eighteenth to the twentieth century. What seemed like cruelty and hardship was just the way life was. Those tough old men were so used to the levels of hardship they endured that they didn't notice: it was all they knew, capitulating to modern comforts was tantamount to surrender.

# 8

## The Present Day

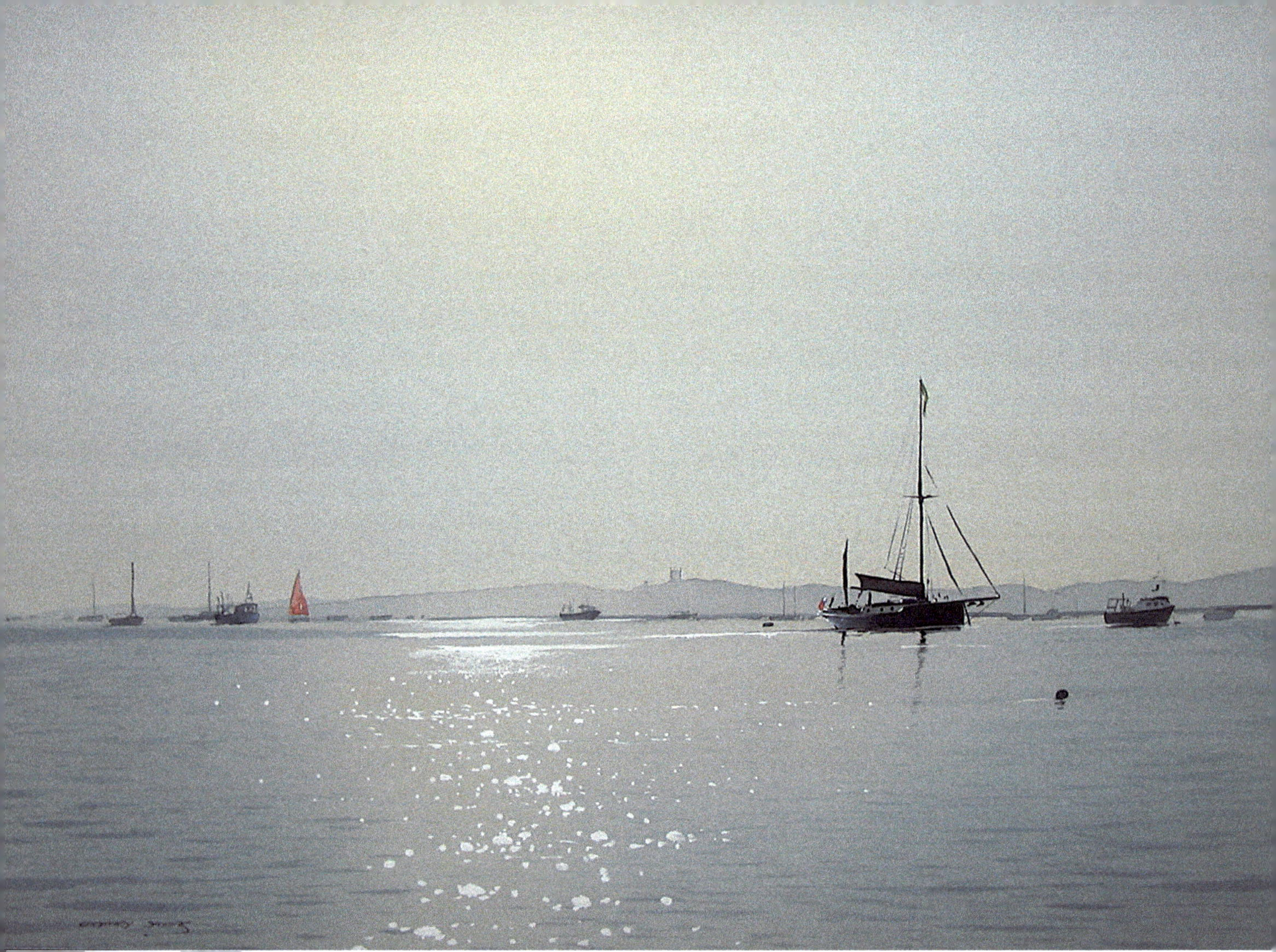

After several months visiting the people and seeing the physical changes that have shaped our small part of the world, we are back in 2020. Apart from the re-dredging of Cley New Cut in the 1980s and some recent minor tinkering, man's physical interventions in the natural processes of the harbour are over. But new fault-lines have developed, growing problems over the futures of the embanked areas and others to related the huge increases in mobility that have brought great demographic change.

During the centuries we have visited, it has been landowners claiming land from the sea that stole the tide. Now it's those same embanked areas, long isolated from natural processes that are becoming a further problem.

From the moment these areas were embanked there has been an increasing disparity between marsh levels, with every passing century these have become increasingly significant. The bank that bounds Blakeney Freshes, built in the first decade of the seventeenth century, has caused them to subside significantly since that time. The saltmarshes have done the reverse, growing with every spring tide.

The row of wooden posts[1] embedded in the marsh. That I mentioned in the introduction they stood two feet above the marsh when I was a child. Today they protrude less than an inch, a clear measure how much the saltmarsh has accreted in sixty years.[2] So it's not difficult to visualise how much they have lifted in 400 years. If you want a clearer picture just walk out along the sea bank until you get to the first sluice gate and look; the difference in levels is startling[3] and this disparity presents a number of problems; to explain them we will begin at Salthouse. The marsh between Salthouse and the sea was the first to be embanked in 1510,

---

1. These posts were part of an assault course for soldiers stationed at Stiffkey camp.
2. Accretion. Every tide, especially storm-driven tides carry a burden of sediment. As it covers the saltmarsh the tide flow eases it drops, thereby raising the marsh level. It is also the very soft mud you can see in the creeks and places like Cley Quay.
3. The difference varies from between two and three metres.

and having been isolated from the natural system for so long makes it the lowest lying of all the embanked marshes. For Salthouse that is problem enough, but there is a compounding factor. The shingle ridge east of Cley Beach carpark was badly mismanaged during the twentieth century. Bulldozed for fifty years into a tall heap, it looked good and acted as a reassuring placebo for those who lived behind it, but for locals who knew better (but were never listened to) it was just a pile of loose stones that breached every time there was a storm-driven spring tide. As the years have passed the naturally formed shingle ridge to the west has stayed in place, over-topping occasionally; what was left after the bulldozers went away hasn't. Had nature provided a longer period of quiescence it might have reformed into a stable ridge. But nature isn't that courteous, so every tidal surge and wind-driven spring tide since then has spread more of its shingle over the marsh behind.

A couple of years ago, a quite normal wind-assisted spring tide put almost five feet of water over the Salthouse to Cley stretch of the A149: the road around Blakeney Quay was covered by just over a foot. As the shingle ridge continues to degrade between Cley and Salthouse so this problem will get worse.

Salthouse marshes have great ecological value and as we move towards Cley so that benefit increases, as do the many wildlife designations that protect it. These are well intentioned and have had remarkable results over the years, but they are not designed to take account of or adjust to geomorphic change. This inflexibility means the embanked marshes form a block in a naturally evolving system and a serious barrier to a sustainable future, both for themselves and the natural functioning of the coast. They cannot be excluded from these processes for much longer.

The situation for Blakeney Freshes is no different. Just as when as a child, on Quay Corner sands I would dig holes as the tide was coming in and pile up sand walls around them to keep it out; it would become a furious battle which could only be lost. A closer analogy to Blakeney Freshes would be hard to find.

A few years ago I was fortunate to be part of the steering group for the latest Shoreline Management Plan Review where all these problems were discussed. And it was here that the fault lines became very clear. The group comprised all the relevant authorities (Environment Agency, Natural England, English Heritage, NCC, NNDC, and the like) but a more perfect example of how democracy can so often fail would have been hard to find. Taking a group of school children to a lecture on quantum physics would not be an unfair analogy. Nobody's fault: the geomorphic processes of this coast are very complex and it would be unreasonable to expect anyone there to have a gained a clear grasp of them within the timeframe allowed. But it made it very difficult to get consensus. The Dutch experts Royal Haskoning gave professional advice but the message they had, which was to allow the tide back into Blakeney Freshes, was, with one obvious exception, unpalatable to all of them. But even if everyone had been in agreement this recommendation would have been very difficult to achieve. Blakeney Freshes is covered with every designation there is, including the all-powerful European Habitats Directive. This meant that had the realignment option been chosen, an equivalent area would have had to be found elsewhere to compensate for the loss. And Blakeney Freshes are so special that that would probably have been impossible. The Environment Agency was left with, quite literally, a narrow strip of firm ground to walk between wildlife and the future functioning of the coast, putting them between the devil and the deep blue sea.

Just months later, in 2013, nature joined the argument and did exactly what Royal Haskoning said it would and Blakeney Freshes were flooded by a tidal surge. This offered a perfect opportunity to allow the tide back in permanently. A tidally functioning Blakeney 'Saltings' (as they would then become) would have doubled the tidal prism of the harbour and guaranteed Blakeney a viable link to the sea for the foreseeable future, along with incalculable economic benefits. But the Environment Agency was tied down like Gulliver; wildlife, the economic value of the Coast Footpath and the general lack of understanding were insurmountable barriers.

So they took the only option open to them. Here, their words outline this:

*'We have concluded from our investigations that the best and most sustainable approach is to repair the damaged embankment with a profile which has a lower and wider crest and shallower slopes. This will be more resilient to damage during any future surge events than the previous embankment design. We also want to reduce the time that saltwater stands on the marshes and we are looking into improvements to the existing sluices and drainage system'.*

What they have done is create a situation where slowly, this area will be eased back into the natural system. Unfortunately, it will be too slow to have any benefit for Blakeney.

The 2013 tidal surge came within inches of disastrous flooding for Cley and Wiveton. Water was running over the Glaven Bank where cows had trodden it down. The Environment Agency immediately acknowledged the seriousness of this, and in a very short space of time raised the bank and reinforced it. But a problem remains; when such a low-lying area as Blakeney Freshes floods in a tidal surge the water is deep enough for the wind to throw up large waves, waves powerful enough to destroy any clay bank in very short order. It is a matter of personal regret to me that the advice of Royal Haskoning was not taken. Sadly we live in an age when, even with good evidence, pragmatic or unpalatable options are frustrated by bureaucratic complexity, and the consequences left to future generations. And this, sadly, seems the most likely outcome here.

Along with the non-governmental organisations, Natural England has ensured that over the last few decades this entire stretch of coast has been smothered with wildlife designations, SAC, SPAs, SSSIs, Ramsar sites, plus the all-powerful European Habitats

Directive. These have seriously restricted what people may do in or near these areas, and have made what in the past would have been routine activities much more complicated. In some cases, they have made these activities impossible. Thus they have become an impediment to wider sustainability and have encouraged a stultifying mindset.

A very basic bit of harbour maintenance cannot now be undertaken without a thorough impact assessment requiring months of work and additional expense. There are of course occasions when this is very proper and necessary. But it is also quite clear to many of us that those within Natural England who have a single-species agenda, and who see all human activity as damaging to this interest, will ensure that this process is always implemented whether it is necessary or not, simply because they cannot stop themselves from seeing local people and their interests as a problem. In recent years things have improved a bit although the constant replacement of personnel means it's a two-steps forward, one-step back process.

In particular, the consultation process that was required to set up the SAC (Special Area of Conservation) designation for this coast. The 'Guiding Principles' included these important words: 'It is essential that owners and occupiers, right holders, local interests, and user groups should be encouraged to participate in the process at the earliest opportunity.' In other words, a participatory decision-making structure involving a wide range of stakeholders. As a consequence, local people were able to make a contribution to Regulation 33 (the Bible for the SAC), thereby forcing English Nature to accept, grudgingly, that local people not only created but also largely maintain this environment. English Nature had to have its arm twisted quite far up its back before it would yield the point, and it was the amount of force needed to achieve this that finally convinced me of the existence of an internal faction with a hidden agenda. Really good naturalists would know the extent to which traditional longshore activities contribute to the natural environment.

Geologically speaking, the North Norfolk coast is a young barrier coast in the early stages of its evolution. Most barrier coasts are fronted by outer reefs and strings of islands which, if allowed to evolve naturally is how this coast will one day be. Barrier coasts and islands are dynamic systems formed by the interaction of wave, wind and tidal energies that erode, transport and deposit sediments. By absorbing the impact of high-energy marine processes, barrier islands reduce the erosion of the mainland coast and provide shelter for sensitive coastal habitats. Eventually, the North Norfolk coast will fit this description, although the shingle ridges and dune systems already do the job very effectively.

The south-east of England is sinking steadily due to post-glacial rebound, and the slow melting of ice during the present interglacial period is raising sea levels. The reclaimed areas along this coast cannot forever be isolated from this process. Subsidence of the reclaimed marsh and centuries of accretion of the saltmarsh have left Blakeney Freshes up to a metre-and-a-half lower than they otherwise would have been. This is quite unsustainable, and one can sensibly argue that the sooner the problem is addressed the easier the solution will be. To do nothing means we either leave a very difficult legacy for our grandchildren, or a natural catastrophe will do it for us. Neither of these is an acceptable option.

Blakeney Freshes is the large area (40 hectares) of freshwater marshes between Blakeney and Cley, which runs round the corner into Wiveton. It constitutes some of the very best habitat of its kind on this coast. Its proximity to Cley Reserve and the Glaven Valley, plus the sea and the saltmarshes to the west, makes it very special indeed, the jewel in the crown of North Norfolk's coastal reserves, one might say. To argue that it should be returned to the sea seems perverse, but like most complicated things there is more to it than meets the eye. Managing this coast clearly requires some very difficult choices to be made in the years ahead, and the 'managed realignment' of Blakeney Freshes is one of the first that will have to be confronted. A decision of this kind will never be easy, and there is a real risk that the many wildlife designations have the potential (if they continue to be used to obstruct) to make such a decision very much more difficult, if not impossible, extending the process until nature loses patience and produces a catastrophic realignment of her own.

Managed realignment, the term used to describe returning embanked land to the sea, was one of eight options set out when the Glaven River had to be diverted a few years ago. Wiveton Parish Council, along with those of Cley and Blakeney, gave an enormous amount of time to the consultation process and many of us argued very strongly for realignment as the most sustainable and forward-thinking option on the board. But it was dismissed. When we asked why, and how the points for the various options had been added up, we were told that the chosen option, that of moving the channel back a couple of hundred yards, was the only one the Department of Environment, Food and Rural Affairs would countenance. (I am recounting this now simply to show that the argument I am about to make has already been rehearsed.)

Why, you must be wondering, can letting such a valuable asset be lost to the sea possibly be the best option? Well, it would not be lost. That is just an example of the language used by opponents. To see why it is the best option requires us to take a much broader view, one that includes local people and a much bigger slice of time.

Blakeney Freshes are very special. If they were to be abandoned, the European Habitats Directive would require Natural England, in conjunction with the National Trust, which owns much of it, to replace it (all of it) elsewhere, although initially they would be expected to try to do that in the immediate vicinity. Some of it (the most important part) could undoubtedly be accommodated within the Glaven Valley. But only some, and because it is so special it would be very difficult to find anywhere else that could provide equivalence. Blakeney Freshes would remain as saltwater marshes, which would still be extremely valuable wetland habitat. Added to the replacement habitat, this would create 80 hectares of nature reserve, double what we have now.

If predictions of increased storminess and sea-level rise are anywhere near accurate, then all the areas of reclaimed marsh along the coast will become both vulnerable and potentially very dangerous for inhabitants. The villages and towns that sit behind sea banks are threatened in a far more sinister way than those that do not. During storm surges, sea banks can breach, sometimes in a catastrophic way, so what in other places might be a steady and predictable rise of the tide becomes a tidal wave. Streets can go from being rain wet to being under four metres of water in as many minutes. This is the deadly threat to those living behind sea banks.

Blakeney and villages like it, which are fronted by dune systems and saltmarsh, are much less exposed to these dangers. The dunes reduce wind, and the much higher marsh levels dissipate most of the wave energy. Allowing Blakeney Freshes and all the other parcels of reclaimed marsh along this coast to go back to the sea would enable it to evolve naturally as a barrier coast, and at the same time extend the same ongoing protection that Blakeney enjoys to all communities currently at risk. Using wildlife designations to hang on to wildlife sites for as long as possible presents a major threat to human life.

Blakeney Freshes comprises grazing marsh, reed beds and some low grassy hills that I believe were sand dunes a thousand years ago. I believe this because one of my terriers once dug deep into one and scraped out nothing but fine, soft white sand. Over the 400 or so years since this area was embanked, much has changed. The surrounding saltmarsh has continued to accrete, growing imperceptibly with every spring tide that covered it, and is now several feet higher than the fresh marsh, which has subsided. It is this, and the possibility that saltmarsh accretion may accelerate with sea-level rise, that makes keeping it untenable. This of course is the empirical view. But there is another view – less pragmatic, more romantic perhaps – but no less important from a human perspective. We each have our own relationship with the world around us, and our happiness can have much to do with the extent to which we are in harmony with it. How beautiful we find it, how much we enjoy it through our work or play, and, perhaps most important, through our association with it over time. Landscape beauty would certainly be enhanced by realignment, but in the short term people's attachment to the landscape could be one of the great obstacles. Many local people, particularly the older generation, those who have lived here all their lives, may not want it to change. Letting the sea back in will wash away more than just paper designations. It will wash away memories. Parts of people's lives will disappear, and that would be felt as loss. The young, I am sure, will see its potential and be excited by it but the rest may have to be persuaded. So important do I believe realignment to be that I want to try and do just that. From my gallery on the Carnser at Blakeney, I regularly saw families pull onto the carpark, get out of their cars and, with the children running, make their way over and up onto the sea bank expecting to find the sea beyond. They are always surprised and disappointed that it isn't there. Although technically dry land finishes where the spring tides leave their tangle along the quayside, there is an awful lot of land between Blakeney and the open sea.

What if those children could run up the bank and not be disappointed? What if they were greeted by a breath-taking view, a vast blue stretch of water as far as the eye could see, great saltwater lakes dotted with islands and surrounded by vast mudflats and saltmarshes? Sailing dinghies out in the distance would be making their way through the old channel, where enough water would remain for it to be navigable even when the tide was out. What might be possible? A small marina for dinghies between Cley and Wiveton where youngsters could learn to sail in safety. A new coast path. Even a road, between Blakeney and Cley, with a string of possibilities all the way along it. A land and seascape that would add magnificently to the beauty of our coast. The benefits for the people of Cley, Blakeney, Morston and all those who visit them would be enormous.

Wildlife would benefit too, waders on the mudflats, geese on the saltings and ground-nesting birds on the islands. Which would also provide great roosting sites. The birds live quite harmoniously with man in the harbour, and they would do so here too. The increased amount of water flowing in and out of the harbour on every tide would scour out the channels, allowing bigger boats once again to reach Blakeney Quay. Blakeney Harbour would have a lease of life that would extend its economic viability for many decades to come.

# 9

## What Might Have Been?

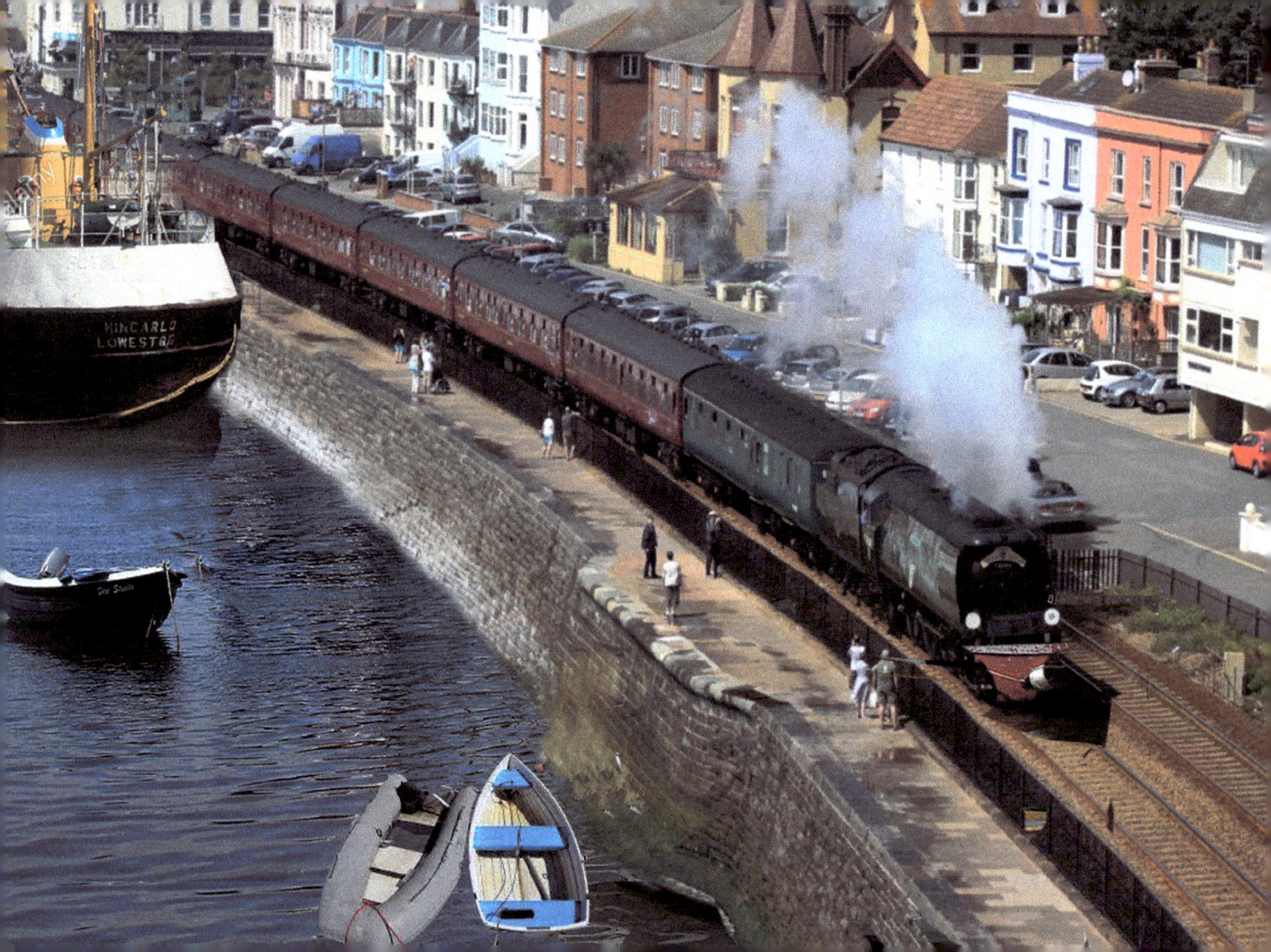

From the very start, men with spades, doing a relatively small amount of digging, set in motion a cascade of natural change that over the centuries would diminish a great Haven into the silted up backwater we have today. They began so early that while the Haven certainly did attain greatness, it never came near to the greatness it might have achieved if someone had locked their tool shed.

So how different would it have been if the ship owners and merchants had wielded the spades and dug out the channels instead of allowing the landowners to block them and build banks to keep out the tide?

The date is 19 February 2020 in a universe several times removed from our own and where in the fourteenth century the balance of power was reversed. We are meeting a local historian who will be our guide and is set to reveal a very different world. We are arriving at Sheringham station where we will change to board the Sheringham-Wells Heritage Railway to take us on to Glaven Port. We are travelling on a coastal railway line that was shut in the 1960s but had been in use for freight and passengers from the 1870s.

For the first part of our journey the coast is reassuringly familiar. Not until we approach Salthouse does the degree of difference we are about to experience begin to dawn: this is a very different world. Salthouse is not a village;

long before we would normally arrive, houses and old converted warehouses appear strung out along the front, to the north the tide is in, a large expanse of water stretching out to the shingle ridge, which is more distant than in our world. Soon there is a quayside on our right with many small boats tied up. As we pull into the station we see our guide waving as the train comes to a stop. Here we will get off and explore Salthouse old town.

The layout is the same as we know now; Cross Street is still Cross Street, but so different, cars parked all the way up, shops and businesses either side until we come to an old flint archway. This leads to a path up to the church. Going through it we are back in the world we know, ahead of us the familiar grand building stands on its grassy hill. Only the larger number of gravestones different. Our guide thinks this is a good place to tell the town's history. He shows us the old ship graffiti and uses the changing ship designs to show how the port evolved.

Salthouse and Kelling, he says, were sheltered anchorages before the villages of Cley, Blakeney and Wiveton (all later to be Glaven Port) came into being, sheltered by the low glacial hills to the north. Both Romans and Vikings sought shelter here and built small settlements. By the fourteenth century, when the spit we know as Blakeney Point had extended its arm to the west of Cley, this place was important, a good deep channel led up from the Haven entrance to the quays. These were further to the north of the quay heading today. All the vessels of the time could reach these quays, but cartage to them over the rough marsh track was difficult as it was so often covered by the tide. The solution was for a smaller creek that ran closer to the town to be dug out into a new wide channel, quite an achievement for the day, but when finished it allowed ships to get right up to the village and

SHIFTING SANDS                                                                                                                                                        145 regular dredging over the centuries has kept a navigable channel to the quay heading you see here today.

This town enjoyed four centuries of prosperity with ships of the day all getting up to its quays, but in the late seventeenth century the relentless landward march of the shingle bank began to pour shingle into the north side of the channel and looked to be sealing Salthouse off from the sea. But it seems that the large volume of water ebbing from the upper Kelling marshes forced its way through the narrower channel, cutting away the marsh on the landward side. At this point the merchants decided to give nature a hand. We have records that tell us how by digging out angled cuts into the eroding bank which caught the outflow, very much more was washed away, and this process sustained over the centuries meant that a navigable channel was maintained right up until the early nineteenth century, when silting limited the size of vessels that could get to the quays. During these prosperous years of trade, Salthouse grew from a small settlement to the busy town we see today, with a tidal channel that is still viable for small yachts and other pleasure craft, making it the perfect place to build the current small-boat marina.

It's time for a coffee, so we make our way back down to the waterfront to a cafe in one of the old granaries. As we sit looking out the window the saltmarsh is beginning to emerge as the tide drops. A couple of small sailers are creeping up against it, struggling as the wind dies. We have arranged to have lunch in Glaven Port, so coffee drunk, we get along to the station. Our historian guide comes with us as he has agreed to accompany us for the entirety of our visit, he also know the best places to eat. This stretch of the heritage railway has two lines and as our train comes in the one we arrived on earlier is returning. How bizarre for us: two trains at a station in Salthouse.

The journey to our next stop and first in Glaven Port is a short one. As we leave Salthouse the landscape around us is familiar, but once past Walsey Hills (called the Old Shore Cliffs in this world) buildings appear, fishing sheds and warehouses to begin with, but soon houses, hotels and other businesses face the tracks on the landward side and the marsh is gradually becoming more exposed on the north. We are heartened as the train pulls in to see that the station name is 'Glaven Port alias Cley', the station seems to be more or less where the mill stands in our continuum, ahead down the track what seems to be a long steel-girder bridge. Our guide tells us it has a swing section over the main channel to allow vessels to get up to Wiveton and Newgate, although he admits it doesn't swing much anymore as both quays are only accessible now to smaller vessels. Coming out from the station we are in a small square, although Plein might be a better word. This doesn't look like Cley at all, or anywhere in England for that matter, it looks more like Flensburg with tall, round, gabled buildings; this, we are told, is the old Dutch quarter of Glaven Port. Moving into town, large three- and four-storey buildings, former granaries and warehouses flank us, now blocks of flats; nothing of the Cley we know exists here. We turn down one of the many narrow alleyways that lead to the estuary. At the quayside we do a brief mental double-take as the view briefly takes us back to what we witnessed in the sixteenth century, a river as wide as the Thames, but seeing the further shore so completely built up we realise that this is a town with a river running through it. Looking north we can see the full length of the railway bridge spanning it, a powerful tide is ebbing from the estuary the sides of which are fringed with saltmarsh. Further upstream, a modern, wide-arched road-bridge spans the whole width, and an island of mud emerges from the tide as between the bridges the channel divides in two, the furthest from us leads to Wiveton the other to Newgate.

Our guide tells us that for centuries the only high tide crossing was over stone bridges at Glavenbridge, inland from where Wiveton is in our world, and that this road bridge was built just sixty years ago to span both channels. Between these bridges are the quays, vessels drawing up to ten feet are still able to reach them, but of course few cargo vessels these days draw so little. Nevertheless, a number of large boats are tied up, fishing trawlers and a couple of replicas of the kind of traders that visited here until the turn of the twentieth century. We are puzzled that some of the old names, like Newgate and the village names, remain while so much else does not. Our guide explains that the two histories ran in tandem until the early sixteenth century when their paths diverged; in our world, the landowners did all they could to reduce the size of the Haven, building banks to enclose land for agriculture, while here the merchants and ship owners stopped them and did everything they could to improve the tidal flow both in and out.

Our guide has arranged for a car to take us through town to see Newgate and on to the next crossing, built during the eighteenth century opposite what once was Wiveton. It feels very odd driving through this busy town, with occasional glimpses between buildings of the other side of town across the water. The name Glandford, our guide tells us, does not exist here; here it was Glavenbridge, but the name only exists now for this road. He adds that the village called Glavenbridge was also further upstream than the Glandford that we know.

We round the corner expecting to see Newgate Green – a surprise, not a green but an area of wet sand and mud dotted with small sailing dinghies the last of the tide running off it. We drive along a low quayside past big buildings toward the church. As it comes into view it takes our breath away. In our time, the money ran out and parts of it lay unfinished. Here in the sixteenth century they had wealth to spare and made it into a cathedral, no small tower tacked onto one corner, here a magnificent spire sitting between two grand transepts, a building to match its image on our 1586 map. The road carries along a curve of quayside backed by grand houses to the other side of the inlet where the church appears even grander, its reflection floated to us over the wet mud. We turn away from the inlet and drive on along Glavenbridge Road; signs direct us left onto an approach road to access the crossing. We rise gently as this is an embankment, a wide road running along the top, to a bridge over the channel at the far end. There is a bit of a wait for traffic but once on the causeway we can see Wiveton church on the other side; it has a strange familiarity, but like Cley it is grander than the one we know.

'But where is the Stone Bridge?' one of us asks.

The explanation, our guide tells us, is complicated. Because the entire Haven was so much larger with more powerful tidal dynamics, the channel at this point was too wide for the technology of the time to bridge, so the medieval bridges were built further upstream at, as you might guess, Glavenbridge.

Time is getting on and we are due to have lunch back in Cley. Coming off the embankment we take a sharp right turn off a roundabout to follow a shore side-road down to the large, new road bridge. The buildings on our left vary a great deal, some as in Cley and Salthouse are converted warehouses and granaries, but there are many modern houses, shops, hotels and restaurants, many with harbour-side areas of seating with parasols and fancy potplants and trees. Across the water an urban landscape, not the countryside we know, just houses and streets up to the hill-top horizon, the frontage dominated by the church with its spire. The channel this side has moorings all the way down with light cruisers and motorboats swinging downtide, and a litter of small dinghies on running moorings along the shore. We take another access road to get onto the bridge and once more have to wait for a break in the traffic, then a further similarity with the Thames, crossing a bridge almost as long as London Bridge. Lunch is a café at the very top of one of the converted granaries, four floors up, and commanding a view of the entire harbour. In the foreground is the railway bridge we will cross this afternoon on the next leg of our journey.

Beyond the wide waterscape that is Glaven Port Harbour, the tide has dropped, exposing the saltmarsh and mud flats and the wide channel running through it, to the north we can see where the Salthouse channel joins forces to add to the massive scouring that has kept this place viable for so long. We note as before that with the shingle ridge more distant there is a greater expanse of marsh, mud and creeks and where we would expect to see the extensive sand dunes of Blakeney Point, just a distant shingle bar with just a few dunes. A Thames sailing barge is creeping up on the last of the tide. It has all its brown sails hoisted but must be using its engine as there is very little wind. This coupled with the smell of the steam train crossing the bridge takes us briefly back to the nineteenth-century heyday of this vibrant town. The train gone, we see the bridge begin to swing to allow the barge to pass. This is a big place, the fourth largest town in the county after King's Lynn, our guide tells us, and in the sixteenth century it was one of the largest in the whole country.

Lunch over, we must be getting back to the station to continue our journey to west Glaven Port. As soon as we are moving, more strangeness: seeing a vast estuary through the passing metal girders of a railway bridge and the fog of smoke from the steam engine is an impossible image to slide over the one we know, add to this the urban world on our left, and the otherness is total.

Slowing as we approach the next station (Glaven Port alias Blakeney), we clatter over the points of a spur line that heads off north to run alongside a wide tarmac roadway leading to what looks like an oil terminal and quay where a number of larger vessels are moored. Cley and Wiveton were definitely for the tourist. This place means business. All the properties along the front are commercial; offices for oil and wind-farm companies, marine equipment suppliers and even a bank and a couple of small factories; this is the working end of town. To accommodate all this the quayside is much larger than any we have seen, between the rail lines and quayside an area full of chains, buoys, fuel drums and equipment of all kinds. On our left a wide road and then the frontage of buildings. Our guide explains that the deep-water moorings (the Pit, in our world) lie inside the curve of the spit closer to Blakeney and that coastal oil tankers moor there to discharge into smaller lighters that bring both oil and gas to the terminals we can see.

We ask why we cannot see the dunes of the Point from here, and the expression on our guide's face suggests another tide-and-time explanation is coming up. The Point 'as we know it', he tells us, evolved abnormally, a shingle spit that moved rapidly west over the centuries with a curve of shingle embracing the harbour in such a way that sand dunes readily developed at its head. Embanking projects so reduced the volume of water ebbing and flowing on each tide that the dunes were able to grow extensively. Here that didn't happen, with as much as four times more water ebbing and flowing, the growth of the Point was kept in check, not reaching as far to the west and also held further out to the north, thereby allowing for a much bigger body of water to serve as a sheltered anchorage.

Our guide takes us along the quay and up the main street to the hotel where we are to stay. When we have settled, he is to come and collect us and take us to a club in town where we are to meet some of the locals and others who work here. The club turns out be more of a nightclub but a quiet spot is found and our guide introduces us to those he has brought to meet us. They are young and old, an elderly man who was born here and has worked all his life on the tugs and lighters of Glaven Port, another who works the oil rigs and a couple more who work for a wind-farm company. The elderly man is a historian like our guide, although unlike our guide he doesn't know where we are from.

'This place has had such colourful history', he says, 'I hardly know where to begin. Up until the mid-seventeenth century, Glaven Port was three separate villages, Cley, Wiveton and Blakeney. You'll have seen some of those names at the railway station. And it was the railway what brought the big change, the three villages had grown and were all prosperous, but the coming of the railway boosted trade no end and quickly they grew into a single town, which was renamed Glaven Port. This place has changed so much in my lifetime', he says. 'When I was a boy steam coasters got up to the quay, as well as Everard's barges, them with barge boards on the side. After the First World War a few steamers still used to come up but they soon got big and the channel was never wide enough for anything of any size. Bigger vessels came into the outer harbour of course but their cargo always had to be brought up by lighters, just as they do today. The port then declined again for a bit until they discovered North Sea oil, that changed everything. Them big companies could afford to bring really big dredgers in to clear the channel making the lower harbour deep enough for the exploration vessels they used, it gave this part of town a new lease of life.'

At this point one of the young men working here chips in.

'He's quite right', he says, 'but the oil business has declined in recent years and that's where we pick up the baton; wind power, that's the future, this harbour is perfectly placed for all the major wind farms off this coast. The heritage railway takes tourist during the day, but you may have noticed that the terminal has a rail link. During the night turbine sections, gear boxes and all the other components for wind turbines come in to be transferred onto lighters that take 'em down to the harbour from where they are transported out to the farms. The income from this has helped keep the heritage line going as well as keeping this harbour open despite the silting.'

These chapters are all extrapolations, stories threaded on to a string of geomorphic change and historic record. Not all of it would be acknowledged by some local historians, the idea that there was a settlement on the end of the spit during the three hundred years to the end of the fifteenth century is still contested, but not all local historians are familiar with the physical dynamics of the coast or the day-to-day details of the fishing industry, which would find such a location perfect for their needs, just as in a later century it was perfect for the Blakeney Lifeboat station. The colourful histories these villages share lie beyond dispute, the evidence for which is not just in books but also all around us, on the marsh, on the beach and in all of the churches. So, put this book down and go and have a look.